MW00444213

Great Communication = Great Production

Great Communication = Great Production

By
Cathy Jameson, M.A.

PennWell Books
PennWell Publishing Company
Tulsa, Oklahoma

We gratefully acknowledge *Dental Economics,*
Dentistry Today, Dental Teamwork, and *Chicago Dental Society Review*
for allowing us to rewrite previously written articles.

Cover photography by John Trammel Photography, Oklahoma City, OK.

PennWell Publishing Company
1421 South Sheridan/P.O. Box 1260
Tulsa, Oklahoma 74101

Jameson, Cathy.
Great communication = great production/Cathy Jameson.
p. cm.
Includes bibliographical references and index.
ISBN 0-87814-427-7
1. Dentistry—Psychological aspects. 2. Communication in dentistry.
3. Dentist and patient. I. Title.
RK53.J35 1994
617.6'023—dc20
94-32276
CIP

Printed in the United States of America

1 2 3 4 5 98 97 96 95 94

TABLE OF CONTENTS

PREFACE

"Creations absolutely de novo *are very rare, if they occur at all; most novelties are only novel combinations of old elements, and the degree of novelty is thus a matter of interpretation."*

SARTON, 1936

I am a teacher. I am proud to be an educator. I am grateful to the people in my life who have taught me the skills of teaching, communicating, writing, and interacting with others. I graduated from the University of Nebraska at Omaha with a degree in education and began teaching in Millard, Nebraska. I had no idea back then that those years of learning how to teach and then actually being in the classroom were preparing me for what I am doing today.

Today, I am doing what I love the most: teaching. But my classroom is different now. My students change every day. My subject matter is business and personnel management and communication, instead of English, math, science, and history.

As a teacher, I am proud when my students—dental professionals throughout the world—learn the skills I am sharing, and when they incorporate those skills into their own lives. I tell my consulting clients that my goal is to teach them so well that at the end of our consultation experience they will not need me anymore. That's what good teachers do. They teach so thoroughly that when the class is finished, the students can function independently. As a teacher, I know when to back away and say, "good job," to my loyal and faithful students. When my students use the communication skills that have been taught and don't remember where or when they learned them, then I know that my purpose has been well served: the students now own the skills and they couldn't go back to the old way—even if they wanted to do so.

I am a teacher. I have had the privilege of learning from others throughout my lifetime. I would like to give my thanks to a few of those teachers who have affected me in a positive way. I have been and will continue to be a student. From my undergraduate days in the seventies to the present, as I complete the final year of my doctorate in psychology, I have had teachers who have influenced me—not only by providing me instruction on fabulous material, but also by providing me with personal guidance. They have believed in me and have helped me to gain confidence in my own self and in my own teaching abilities. To them I say, thank you for being excellent teachers. Thanks for seeing talent in me. Thanks for encouraging me to "step out there" even when I didn't have confidence in myself. Thanks for loving me. May God bless all of you now and forever.

Ron Reichert, Ed.D., who saw me in the classroom as a young, naive undergraduate student and said, "I want that young lady on my team." Thanks for the years of Camelot, Ron.

Bettye Brown and Helen Appelberg, Montessori instructors, who took my teaching abilities one step further and who opened the door for my instruction with Effectiveness Training.

Thomas Gordon, Ph.D., for his incredible gift to mankind through Effectiveness Training, Inc. Thank you for allowing me to be a certified instructor of your information and for having influenced my personal and career life through your own brilliant teaching.

Robert McLaughlin, Ph.D., who advised me throughout my

program of study for my master's degree in psychology and who encouraged me to take the next step: the Ph.D.

Dr. Burt Press and Dr. Jim Saddoris, both former presidents of the American Dental Association, who have taught me skills of management and who have opened doors for me because they believe that what I offer will benefit their beloved profession. Thanks for the support and encouragement.

Steve Emmons, Effectiveness Training, Inc. and IBT—communication, organization, and time management. Tom Hopkins of Tom Hopkins International—communication, presentation, and sales skills. Ed Foreman and Zig Ziglar—goal setting and sales skills. Ken Anbender, Ph.D.—my personal consultant—life design, well-being, corporate organization, and encouragement. Reverend Bill Crawford—prayers and spiritual guidance. And Father Arlen Fowler, who knows more about me than anyone—and loves me anyway! Thanks for the counseling, the teaching, and the lessons on life.

A loving and grateful thanks to my fellow PennWell author, mentor, teacher, and friend, Karen Moawad. Karen, you opened your heart, your mind, and your "consulting materials" to me when I didn't even know what a consultant was! You saw something in me and reached out to me, just as you do with so many people.

I want to be just like you when I grow up! You not only teach, but you touch lives with such unconditional love that everyone who is a recipient of that love grows to heights before unknown. Thank you for your insight, your unselfish sharing of material, your ever-faithful words of wisdom and advice, and your solid friendship.

No book can be written, no teacher can function effectively without a great support staff. Thanks go to Jolene Coffey Smith, Vicki Nashert, and Crissy Bixler for their patience, their loyalty, their belief in our mission and their willingness to get me to those deadlines!

Thanks to my supporting team of consultants: Belinda Mitchell, Dru Halverson, Pam Johnson, Bonnie Peterson, and Dr. John Jameson.

With love, thank you Brett and Carrie—what a gift I have received to be your Mom! Thanks, Dad—for teaching me about commitment to work and for the "drive." Thanks, Mom—for being a role model whose softness and patience offset some of my "fiery ways." Even though you are gone from us, you shine through all of us—every day. Thanks, MiMi—people ask how I do everything I do. I can answer with one word, MiMi. I am forever grateful. Thanks for the ever-willing "helping hand."

And last, but not least, thanks to John for being my best supporter, for believing in me, for always knowing when I need an emotional boost or a cup of coffee or a big hug. Thanks for "allowing me to become." Thanks for being a role model for me, for our teams, for our community, for our profession, and for our family. I ALWAYS know that I can count on you and that you will be there. You're the best. I love you.

I give thanks to God for His strength, His wisdom, His guidance, His support. Every day I pray, "If someone here today needs to hear Your word, let me be a vehicle for You. If someone needs Your message, may I have the privilege of being Your messenger. I am nothing. You are everything. I pray that I will be about Your work every day of my life and that my work will be pleasing in Your sight.

"I know that through You, Dear God, all things are possible. I dedicate this book—and all of my work—to You. Thanks for being there. I need You and love You."

CHAPTER 1

INTRODUCTION

"Believe that there is genuine magic in believing—and magic there will be. For belief will supply the power which will enable you to succeed in everything you undertake. Back your belief with a resolute will, and you become unconquerable—a master of men among men—yourself."

CLAUDE BRISTOL

As the wife of a practicing dentist, I have long been concerned about and interested in dealing with the day-to-day challenges that my husband John faces in his own practice. As a result of being actively involved with John's dental school education and then with his practice for 20 years, I have seen, lived, and dealt with the challenges of the dental profession and of running a business.

Now, as a teacher to hundreds of dental practices across the country and as a lecturer for thousands of dental professionals throughout the world, I have become vividly aware of the fact that John's challenges are a reflection of what goes on *every day* in dental practices.

Throughout our years in dentistry, John and I have found that my background in education and communication skills has been an asset to our practice. We have built the business aspects of our practice on a solid foundation of educative and communicative expertise. Our success has been based on a commitment to educate our patients about the benefits of dental care and to communicate those benefits effectively. This commitment to excellent communication, education, and practice management has allowed John to apply the excellent clinical skills that he loves so much.

It is this belief that communication skills allow dental teams to provide the technical skills and that great communication does, in fact, lead to great production that has stimulated my interest in preparing this text.

Throughout this book, I will be referring to "the dental professional" and to the dental "team." When I use either of these terms, I will be referring to *everyone* on the team. Dr. Jim Saddoris, past president of the American Dental Association, says that when he refers to the dentist, he means the entire team: doctor, hygienist, and auxiliary. I love the insight of Dr. Saddoris. In order to get the message across and to provide total comprehensive dental care—physically and psychologically—all members of the team are valuable and greatly needed.

The communication skills taught within the pages of this book will be appropriate for the building of your team, as well as for the enhancement of your patient relationships. Without excellent communication among team members, patient care cannot be at its best.

> *My definition of a dental team is a group of leaders working toward a specific set of goals that leads to the accomplishment of the ultimate purpose or mission of the practice. Each team member performs their specific responsibilities excellently with commitment to the practice, to the other team members, to the patients, and to themselves.*

THE PURPOSE

Dental professionals, working as a team, have as their purpose the care of the oral cavity—a vital organ of the body that affects the entire body. The dental professional is responsible for educating people about the need for excellent care of the oral cavity, with the goal being optimal health and well-being.

The oral cavity is an organ of the body that not only affects all other systems—but it also uniquely reflects attitude and emotion. It is considered an intimate zone of one's body, protected and shared cautiously. Through the mouth, words of conversation expressing thought and emotion are transmitted. The oral cavity is one's vehicle for communication, verbally and physically. Thus the dental professional is commissioned to care for one of the human being's most psychologically expressive avenues: the oral cavity.

This summation serves to express the value of the service provided by dental professionals. You, as a dental professional, must determine and care for a person's physical needs through careful diagnosis and treatment. You must be able to educate a person as to the value of recommended treatment or therapy for the oral cavity. You must be able to communicate effectively all of the above, and you must be able to deal with a person's psychological needs as they relate to problems or issues involving this organ.

In order to satisfy all of these requirements, the dental team needs to be well-skilled in the art and science of effective communication. Without effective communication skills, none of the previously mentioned challenges can be fully met and fully satisfied. If these challenges are not satisfied, the practice will suffer from low productivity, and the professionals may suffer from poor self-esteem. In addition, stress can get out of control because conflicts between team members and with patients develop as a result of frustration and anxiety.

Four Areas of "Need Development" for the Dental Practice

Four strong areas of "need" have emerged through my years of research and work within the dental profession. These prominent themes are:

1. Stress management: Dentistry has the number one suicide rate among professionals. Dentistry ranks as one of the leaders in drug abuse, alcoholism, divorce, and professional dropout.

What leads to this extreme stress? What can be done to relieve—or control—this stress? Can effective communication help to alleviate some of this uncontrolled stress? Can excellent communication skills lead to resolution of problems that might otherwise be causing stress?

2. Staff fulfillment: The turnover in dentistry among auxiliaries is an issue of concern for the industry and for individuals. Turnover of auxiliaries is costly, both psychologically and financially. Dr. Burt Press indicates that every time a doctor loses a key team player that this turnover costs the doctor approximately one year's salary.

What leads to dissatisfaction among dental auxiliary? Can effective communication systems be developed that will serve as vehicles for personal growth and satisfaction between and among dental professionals? Will the end result be less turnover?

3. Practice management skills: Dentistry is a health-care profession, with the service of people as its main purpose. However, dentistry is also a business, and unless the doctor and his/her team are able to manage their practice as a smooth running and *profitable* business, they will not be able to remain in business. Thus, they will not be able to serve people.

One of the management systems within the practice is the system of case presentation/treatment acceptance. A barrier to optimum success and fulfillment within the dental industry seems to be the inability to gain case acceptance. Most dentists have more dentistry sitting in the charts waiting to be done than they have ever done in their practicing years. This undone dentistry can lead to financial stress and professional dissatisfaction. Therefore, one of the most important management systems within the practice is the system of case presentation and treatment acceptance. This seems to be the system that gets the least attention. Communication skills are not taught in dental school, and there are very few courses given or dental books written on the subject of communication skills as they apply to care presentation. Thus, this book.

4. Communication skills: I am thoroughly convinced that communication is the bottom line to success within a dental practice today. A person's/patient's willingness

1. to become involved with a practice or organization,

2. to say yes to the purchase of a product or service,
3. to remain active and loyal to that practice,
4. to refer others to that practice

is a direct result of the team's ability to give and to receive information clearly and accurately. Thus personal, professional, and financial success depends upon effective communication. Indeed, great communication equals great production.

Communication? What Are the Benefits?

This study of communication will benefit you in the following ways:

1. Excellent management skills and systems will be enhanced by improved communication.
2. Quality relationships among team members will develop when effective communication and effective problem-solving skills are learned and used.
3. Effective communication skills lead toward enhanced relationships with patients, which will lead to increased revenues for the practice. People will accept your treatment recommendations, stay with you (no more falling through the cracks), and refer others to you.
4. Fulfillment and satisfaction for the team members will result and you will kindle or rekindle your love for your profession.
5. Stress will be controlled.

WHAT IS COMMUNICATION?

John D. Rockefeller once said that 85% of success is related to people skills and that 15% of success is related to technical skills. Think of your relationships with your patients. Is the success of your relationships centered around your excellent dentistry or around the level of care with which you surround the patient?

Most of you probably answered the question by saying something like, "The patient seems to notice everything except the dentistry. They want and expect the treatment to be excellent, but everything else seems to have a stronger impact." Without question, the clinical dentistry must be superb. However, most of you would probably agree that patients make decisions based on just about everything else besides the dentistry.

Doesn't it make sense to be excellent at both? Can you combine quality clinical skills and a commitment to comprehensive, dentistry with a coinciding commitment to excellent communication? Will both you and the patient benefit as a result of this dual commitment? I think so.

How you communicate, how you relate to the patients, what kind of personal service you provide will affect whether or not a person will accept treatment, schedule an appointment, agree to the financial arrangement, refer others to you, and so forth.

However, even though most people will agree that communication is the bottom line of success, very few people undertake an active study of the skills. Communication is often taken for granted. You might be saying, "Some people are just born communicators." No, they aren't. Communication is a skill. And, because it is a skill—or a set of skills—it can be studied and learned.

Communication is a two-way process in which each party is responsible for the success of the communication. Communication is dynamic, not static. Communication is a mutual transaction between parties. Effective communication means that when a person sends or receives information that a clear or accurate understanding of the message results.

How many times do you hear what a person is saying but totally misinterpret the meaning? And vice versa, how many times do you send a message only to have it misunderstood? Did you not send a clear message? Did you not have the ability to get your point across? Or did the receiver of the message get too hung up in his/her own feelings to give you the attention necessary for an accurate interpretation? All of these are relevant questions when learning how to communicate more effectively.

That's why you have selected this book. You know that the success of your relationships, the success of your practice, and the success of your career balance on your ability to communicate. You are making a conscious decision to get better at those skills.

Follow the teaching within this book. Practice the skills. Incorporate these new skills into your every day and you will see incredible growth. I have. I have followed the teaching of my own mentors, my own instructors on educating, communicating, and presenting, and have found that my personal and professional life have benefited from the study. You will receive the same results. Study hard. Practice with intention. Enjoy the process.

CHAPTER 2

GOAL ACCOMPLISHMENT

"When you have achievable goals
for every day written down, every day
becomes an exciting contract with yourself.
You get up in the morning
with a plan for making that day contribute
the most that it possibly can to getting
what you want from life."

TOM HOPKINS

I cannot start a consultation, a lecture, or this book without laying what I consider to be the foundation for success: *goal setting*. Many similar threads run through the lives of successful people and through successful businesses, but the most common thread seems to be that these achievers have *written goals*—a plan for their own success.

What Is a Goal?

A goal is defined by Webster as "the end result toward which effort is directed." A goal gives you a clear vision of the result of specific actions or behaviors. Goal definition helps to clarify the reason behind each of your day-to-day activities. You will see more clearly that each and every day—and the activities of those days—are stepping-stones to the goals that you have set. This process is motivational. It gets you up in the morning, keeps you going throughout the day and sometimes, into the night. Goals become that guiding light that pulls you through the "thick and the thin" of it all.

Many people have a vague idea of what they want to achieve in life, but very few spend the time and the energy to plan for successful achievement of goals.

A goal does not become a goal until it is written down. Unwritten, a goal is only a dream or a wish. It is okay to dream and to wish. Do so! But take the dreams that mean the most to you and work at turning those dreams into realities.

Goal-Setting Studies

Harvard University and Dr. David McClelland have spent a great deal of research, time, and money on the study of motivation. In studying motivation, they wanted to find what major attributes characterize successful people.

They found that one of the key characteristics of successful, highly motivated people was that these people were goal oriented. In fact, Harvard found that 80% of the productivity in the United States is accomplished by 20% of the population. In analyzing this productive 20%, the researchers discovered that this group was, in fact, goal oriented, but that only 3% of the group *wrote down* their goals.

Expanded study of these people uncovered the fact that the 3% that were goal writers managed to increase their own productivity by a minimum of 10% and their personal incomes by approximately 100% within a year from the time they began writing their goals.

Yale University performed a similar study. They surveyed a class of graduating seniors and in that survey asked if the individuals wrote down their goals. Again, 3% said "yes." (3%! Remember: Different university, different researchers, different subjects! Same result!)

The university tracked this group of graduates for 20 years. At the end of that 20-year period, they found that the 3% of the graduates who had written their goals (and who had continued to do so) had accomplished more in measurable entities (including financial success) than the other 97% of the graduates put together!

On a personal note, we began writing our goals at a time in our lives when we needed to grow personally and professionally. We had been very successful in our dental practice and in other businesses, but the oil crisis hit Oklahoma, and what we had been doing wasn't working anymore. So we thought, "Why keep doing things the same way?"

We had the privilege of hearing a tremendous motivational speaker and management expert, Mr. Ed Foreman of Dallas, TX. Mr. Foreman was addressing a group of dental professionals at the Dallas Midwinter Meeting and was encouraging us to become more effective managers of our lives and of our businesses. He encouraged us to get control of our lives and start doing that by writing down goals.

We decided that the whole concept made too much sense not to do it! And besides, it couldn't hurt. We began to study the art and science of goal setting and goal accomplishment. I wanted to know why only 3% of the population writes goals and commits to the process if such tremendous results can be achieved. I wanted to know about goal accomplishment personally because we didn't just want a 10% growth in our practice, we needed it!

For the past decade and a half, I have studied everything I could find on goal writing, goal setting, goal accomplishment. We have incorporated this process into our professional lives and into our personal lives. The process works better than I had ever expected. The key is getting the mind in the right place and believing that each and every worthwhile goal set and processed can be accomplished.

"Whatever the mind of man can conceive and believe, so shall he achieve."

Napoleon Hill

Read this chapter on goal setting with an open mind. Understand the value that this process can add to your life. Take the first step to gaining control of your life, the first step on the road to success: goal setting.

The First Step

That first step towards change, growth, and success—personally and professionally—is to write the goals. Within a year from the time we began writing our own goals, we had increased the productivity of our practice by 35.5%, and we had, in fact, increased our personal income by about 100%.

We had begun a journey of continued growth and prosperity, a journey to personal fulfillment and to happiness that continues today.

We still write down our goals—and will do so forever. We have learned that getting ourselves and our team focused on a common set of goals and working together toward the accomplishment of those goals provides the guidance, the motivation, and the focus needed for continuous progress. *We gave up the things in our lives that were not working and have continued to do the things that are working*. Goal setting was and is the foundation of our success process.

REASONS PEOPLE DO NOT WRITE THEIR GOALS

You may be saying to yourself, "If this goal-setting business is so great, then why don't more people do it? If the success rate of the 3% is so tangible, why don't more people take this step?"

Good question!

There are several reasons why people choose not to write their goals:

1. Fear of failure
2. Low feeling of self-worth, lack of confidence, low self-esteem

3. Never taught how to write goals
4. Don't see any reason to go through the process
5. Don't know what their goals are

Fear of Failure

How in the world does fear of failure relate to goal setting, you ask? Many people hesitate to write a goal down in black and white, because if the goal is not accomplished, then failure will be seen as the result. "If I write it down and don't get it done, then I will be viewed as a failure." People fear failure, because they see it as a defeat, as a negative. "If I'm not quite sure of myself anyway, then to fail at goal setting will just solidify my perception that I'm not so great!"

The dental professional is often caught in the trap of believing that he/she must be perfect. The environment in which the dentist performs is extremely small and there is no margin for error. The dentist believes that he/she must meet the needs of all people, relate exceptionally well to all people, be a top-notch leader of the dental team, be the total provider for the team and for family, and must not show any emotions that might be misconstrued as negative.

Robert Eliot, M.D., in his comprehensive study of stress management has found that, "Perfectionism is expectation that never meets reality," and it is fueled by the fear of failure. It is comprised of guilt, defensiveness, and the fear of ridicule. Clues to this behavior are the words *should* and *have to.* Perfectionists believe and practice the adage, "If you want something done right, do it yourself." Unable to delegate even the most minor tasks, they become angry with themselves or others whenever any detail can't be done "just right."

This *perfectionist misconception* leads to incredible stress within the dental professional—usually, but not always, self-induced. To write down goals, to put these down on paper, to expose them to one's conscious self or to others is viewed as a possible opening for others to see the humanism and reality and imperfection that might result if these goals are not obtained. And so, it is safer to leave the goals that one dreams of or desires tucked away in the safety of the subconscious self.

The only way to overcome a fear is to face that fear head-on. When we face that fear, deal with it, and survive, we take great strides in truly overcoming the fear. Address fear by doing exactly what you fear the most. Once you discover that you *can* do something and survive, the task will be more comfortable for you the next time.

Facing a fear takes a great deal of courage. Thomas Gordon, Ph.D., says that, "Courage is not the absence of fear, but the willingness to act in spite of fear." The fact that you have faced something you have previously feared is a success within itself. Not ever facing an issue or a task that you fear takes away from your very being. This stifles your journey towards self-fulfillment.

In evaluating the fear of failure as it relates to goal accomplishment, let me propose a different view of the matter. Instead of looking at *not* accomplishing a written goal as a failure, look at it as *"new learning."* You have just learned that the plan you have been following does not work for the accomplishment of a particular goal. You now know that you need to step back, take a new look, develop another plan, and come at the accomplishment of your goal in another way.

The only failure that relates to goal setting is this: if you learn and understand the benefits of goal setting and never even try, then I think you have failed: failed to give yourself the opportunity to be all that you can be.

Low Self-Esteem

Some of you are not overwhelmed by the issue of fear of failure, but you have decided for yourself that you are not worthy of success or the achievement of a particular set of goals. Your fear may not be fear of failure, but rather fear of success. You have told yourself so many times that "I can't," or "I wouldn't be able to," or "I couldn't possibly," or "I'm not worthy" that your subconscious mind has come to believe you.

The subconscious mind doesn't know the difference between reality and nonreality. And so, if your subconscious mind constantly receives messages about your lack of self-worth, it begins to believe that this is true. Efforts to feed your mind positive input about yourself, your abilities, and your desires or goals are essential for growth. You can begin to overcome a feeling of low self-worth by feeding positive thoughts and nutritional information into your mind.

Thomas Edison: Great Attitude

Long ago Thomas Edison (considered one of the greatest inventors in history) was working on the acquisition of a light source. He tried and failed in many of his attempts. In fact, he developed over 10,000 experiments on this light source that did not work! One day a friend asked Mr. Edison if he hadn't become discouraged after failing over 10,000 times in his efforts. Mr. Edison replied, "Failed 10,000 times? I didn't fail 10,000 times. I simply learned 10,000 ways *not* to make a light bulb!"

What a way to look at a situation! He saw the positive value of the learning rather than the negative effects of perceived failure. It's not what happens to you that makes the difference. It's what you learn from the event.

If you set a goal and you find that your plan of action doesn't work for you, don't quit! Learn from your experience. Turn negatives into positives and your life will be a constant learning experience. How exciting that will be!

Don't Know How

Most people have never had a course in goal setting, goal writing, and goal accomplishment. If goal writing is so valuable for success in life, shouldn't this course of study be offered as a course in school, perhaps a required course? But then, success in life is not required, is it? We choose to be successful—and we choose to be unsuccessful!

> *"Destiny is not a matter of chance: it is a matter of choice."*
>
> AUTHOR UNKNOWN

But since you are reading this book, you have made a statement about your desire to get in control of your dental practice, and you have made a statement that you desire to be successful.

And so, I am going to teach you how to go about the business of writing goals. Do you remember that one of the main reasons that people do not write their goals is that they don't know how? Well, at the end of this chapter you will have had a thorough lesson on goal setting—and not knowing how will not be a legitimate excuse anymore!

You may not have all the information you want or need on goal writing at the end of this chapter, so I encourage you to seek other books on the subject and discover all you can about this key to success.

Don't See Why

Remember, a dream does not become a *goal* until it is written down. You may be asking, "Why do I have to write it down? I know what I want!"

Writing the goal down serves several purposes:

1. It then becomes something you can visualize. It becomes tangible.
2. It becomes something you can refer to for evaluation and any necessary modifications.
3. Evaluating your path towards the reaching of the goal lends a sense of gratification and positive reinforcement for work well done.

Educators and psychologists have told us for decades that the key to solidifying behavior is to positively reinforce that behavior. When you take a step forward on your journey to success, when you take a step towards the accomplishment of a goal, pat yourself on the back for that forward step. This reinforces your efforts.

In working on your self-esteem (and you must constantly work on it), this reinforcement delivers a sense of satisfaction that you *can* do something and that your efforts have been fruitful. You see the results of your efforts and are encouraged to continue to put forth effort. In other words—you stay motivated to continue, and you stay motivated enthusiastically!

Don't Know What They Are

When you begin writing your goals, allow your imagination to be set free. Put your every dream down on paper for the purpose of visualizing and analyzing each dream. If you don't think you have any dreams, think again. Reach down deep into your gut and pull out any passions that you may possess and write them down. You may have been indoctrinated to put away your dreams, not to bother with such trivia. Most people have been taught to do just that—and have been taught very well!

Now I'm recommending that you "pop the top" and let those desires bubble out of your inner self.

When writing goals, consider writing them in three distinct, yet interrelated, areas of your life:

1. Personal and family goals
2. Business and career goals
3. Self-improvement goals

The three areas are so interrelated that it is difficult to separate them. Your degree of wellness—your physical, emotional, spiritual, and mental fortitude—definitely affects your work which in turn affects how you feel about yourself, which affects your family life, and so on. There is a sense of connectedness between the three areas.

Make a supreme effort to keep your goals and your life in balance and prioritize those goals. Let nothing get in the way of maintaining that prioritized balance.

Zig Ziglar teaches that in each of these three areas of goal definition, you need to generate thoughts on what it is you want to *do,* who and what you want to *be,* and what you want or desire to *have.* Once you have set your imagination free, once you have reached down deep into your gut to pull out your innermost feelings, thoughts, and desires, once you have written down what it is you want to do, be, and have in each of the three areas of need, then it is time to begin putting a process of accomplishment into motion. Fig. 2–1 shows a format for goal writing in the three areas of your life where goal writing is appropriate. Use this as a springboard for goal writing. If you are a person who does not at this time know what your goals are, this may help you get started.

THE GOAL-WRITING PROCESS

1. Writing the goal: Here is the first step to accomplishing the goal, but it is not the only step! Just as visualizing the end result of a crown preparation does not make the preparation happen, neither does simply writing the goal down on paper make it happen. Just as you have a specific process for that crown preparation, you must have a specific process for goal accomplishment. In the process of goal accomplishment, that first step is to write it down. That's like writing a contract

Your Personal Dream List

"A goal is a dream that is better defined."
DR. T. WARREN CENTER

Personal/Family

Children	Car
Spouse	Vacation
Home	Money
Business/Career	**Self-Improvement**
Salary	Physical
Benefits	
Career Development (Education)	Mental
Equipment	
Financial Security	Spiritual

Figure 2–1 This form is a springboard for writing your goals.

with yourself. Decide what it is you want, and begin the process of accomplishment by writing it down.

In writing your goals, be very specific. The mind needs a very clear picture of what it is that you want to accomplish. So be specific, not general. Be generous in the visual details that you write down and that you play and replay in your mind.

Then prioritize your goals so that you don't get what my good friend Sally McKenzie calls "priority confusion." Some people spend all of their time on the things that don't make the big difference and never get to the significant tasks of their days—or of their lives.

2. Design the plan: Determine what the objectives or strategies necessary for the accomplishment of the goal will be. What must you do to reach this goal? What is the action plan?

Identify necessary resources. What people do you need to consult? What books do you need to read? What courses do you need to attend? What tapes do you need to listen to?

Also identify barriers that might get in the way of the accomplishment of the goal. Then figure out how you are going to overcome those barriers. The first step in solving a problem is to define it. Once you have clearly defined the problem, you can begin to solve it. If a barrier exists that might prevent you from accomplishing your goal, then carefully identify that barrier so that you can begin to chip away at it. Work at eliminating barriers so that there are no walls in the way of your success.

Brainstorming is appropriate here. Brainstorming means that all ideas are shared in a very accepting environment. In other words, you may not like someone's idea, but you are not going to rip it to shreds. Acceptance of another person's view does not mean you agree, it simply means that you respect the other person enough to hear their view or to hear their idea.

Once all ideas have been thrown out and written down, go back and discuss each one—the pros and the cons of each. Come to a consensus agreement as to which idea or ideas you believe will best help you accomplish the goal.

3. Assign the person or persons responsible: Define who will perform or carry out each task, each action step. If a person on the team has been actively involved with defining, writing, and planning for the team goals, that person will respond with enthusiasm when given the responsibility to perform related tasks. If a person on your team knows what to do, how to do it, and why to do it, they will usually perform excellently, and things will get done in an expedient manner.

4. Define the time frame for each step: It is essential that time frames be specifically outlined and recorded. This will allow all persons to hold themselves accountable. The doctor can hold each person accountable as well. This wonderful step of the goal accomplishment process offsets that dreaded disease *procrastination!*

5. Evaluate: On a specific, predetermined, regular basis, analyze how things are going. Are you on or off course? Do you need to adjust your plan? What have you learned from the proposed plan of action? Evaluation is as critical as *any* other step of the process.

After you have designed the specific plan, *put the plan into action!* So many practices fall short right there! They know what they want, they design a great plan, but they never expend the energies necessary to put the plan into action.

MY FATHER, THE ARCHITECT

My father is a brilliant architect whose artistic talents and masterful knowledge of science and engineering have become a statement of who he is. He is one of those extraordinary people who leaves an indelible mark on history and on the lives of those people he touches with his work and his love of it.

I often think of his profession of architecture and building when I am teaching the skills of goal setting.

The end result of one of his designs is a beautiful, functional structure serving those for whom the edifice was constructed. When the building is completed, a celebration usually takes place to inaugurate the facility and to share appropriate accolades with the builders, the engineers, and the architect.

But the process of getting to this point is long and tedious. The process begins many months—perhaps years—before the celebration of the arrival. The process begins with an idea, a goal. Then very detailed, specific plans are drawn and written, describing to the nth degree what must happen before the structure is completed and the goal is accomplished.

The goal is set, the plan is designed, and then the plan is put into action. Work begins. Specific objectives and strategies are assigned to specific people. Everyone knows their responsibilities and is held accountable for the accomplishment of those tasks. A time frame is set for the accomplishment of each task so that the goal of a finished building is met in time to satisfy everyone. Evaluation occurs everyday. How did this fit in with that? How will this portion of the building process prepare for the next phase? Are they on schedule? Are they heading in the right direction, or do they need to make adjustments in the original plan?

And then the building is completed. The goal is accomplished. Celebration is in order. And it is good.

But I have learned a great lesson of life from this great architect: the joy does not come from the end result as much as it comes from the process of doing. The true joy and the ultimate reward is the process of creating: the planning, the building, the growing, the journey on the way to the celebration.

Here's the process:

1. Write the goal.
2. Determine the objectives and strategies.
3. Assign responsibilities to the appropriate person or persons.
4. Time activate every task.
5. Evaluate.

A GIFT YOU GIVE TO YOURSELF

It has been said that many people spend more time planning their vacation than they spend planning their life, and I believe this is true for the most part. We must know where it is we want to go if we are ever to get there! So many people spend an entire lifetime wandering around and then looking back and wondering where life went or wishing they had done _____! I don't know about you, but I want to look back on my life and say, "I'm glad I did _____!"

Design a plan for your life, and then be about the business of detailing that plan. Become an architect in the structure of your life. Start with a 20-year forecast and work backwards: 10 years, five years, one year, monthly, weekly, daily. A whole bunch of well-managed days make for great weeks, which lead into satisfactory months, excellent years, and a goal-oriented, goal-satisfied lifetime.

Setting goals for a lifetime seems somewhat overwhelming, I know. But don't think that your goals cannot or will not change. Keeping your goals written down keeps you on track and allows you to make those very necessary changes that evolve as you evolve. Set long-term and short-term goals. A clearly defined set of short-term goals will be necessary for the accomplishment of your long-term goals. Those lifetime goals may be somewhat vague, but the more specific short-term goals are more action oriented and will assure you that your footing is solid.

For example, you may have set goals such as your desirable retirement age, your desired financial status at that retirement age, and your career goals to be reached by retirement age. But in order to reach those faraway goals, you must set very specific short-term goals, such as (1) yearly production/collection goals, (2) monthly debt-servicing schedules, (3) business plans for career accomplishments per year, month, week, day.

Fig. 2–2 shows a format for goal writing that we use in our office on a regular basis. We follow the goal scenario for everything we do in our office and in our personal lives. We consider the writing and reaching of well-planned goals a gift—a gift that allows us to stretch to reach our individual and group potentials.

Team Goals

Every January our team writes individual and team goals. We encourage the sharing of each individual team member's business and career goals. We want to know what each individual hopes to accomplish personally as a dental professional, and we want to know what

Goals and Objectives

Goal			
Objective	Responsible Person	Time Frame	Evaluation

Figure 2–2 Apply this five-step goal accomplishment process.

they hope to bring to the practice and what they hope to receive from the practice. As the leader of his organization (or practice), John can do a better job of leading if he knows what it is the individual members of his team want to accomplish in the profession and in the practice. If the individual members of the team are working toward their own predetermined goals, who wins? Right! Everyone! Doctor, team member, and patient.

Then we coagulate the individual goals into a set of team goals. These team goals are written down, a plan is designed, specific tasks are assigned to specific persons with a time for accomplishment attached. We put the plan into action and constantly evaluate our progress. At the end of the year, we are constantly amazed and overjoyed to find upon

This Year's Goals

Personal Career Goals: What I want to accomplish in my position in the practice this year.

____

____

____

____

____

____

Team Goals: What I would like to see the team accomplish this year. ____

____

____

____

____

____

____

Figure 2–3 Record and evaluate your individual and team goals.

final evaluation, that most of those goals have been accomplished.

As we evaluate the goals, we are continually amazed at how many of them are achieved in a very short period of time. By constantly evaluating, necessary adjustments can be made before it is too late, or before too much time passes without result. We write new goals once or twice a year with additions, deletions, and adjustments made when necessary or beneficial.

We are forever committed to encouraging each person on our dental team to reach their own unique potential. Goal setting has taken us a long way towards maintaining that commitment. We find great joy in sharing the success of our practice. And when I say "our" practice, I mean just that! The members of our team are co-owners of the practice. They are challenged by the trusted responsibility attached to goal writing and goal accomplishment and are rewarded for individual and group achievement.

We are, indeed, in control of our destiny. We are responsible for our success or our failure. We have chosen to succeed. When you *decide* to be successful, you will be. Until you make the commitment to control your life, you never will.

IN SUMMARY

The first step to success is *deciding* to be a success. The next step is to gain control of your life. You begin that process when you *write your goals!* I hope that you will work on any fear of failure that may have prevented you from writing your goals prior to this reading. I hope you will see yourself as a beautiful, worthy human being with wonderful opportunities available to you in your practice of dentistry. I hope that you now see a reason to write your goals and that you have a better understanding of how to go about that writing. But most of all, I hope that you will start *now* to begin your journey to success and happiness.

Write the word *procrastinate* on a piece of paper, light a match, and burn the paper! Procrastination never helped anyone accomplish goals! Start now! I believe in you: believe in yourself! Be the *best* you can possibly be—now and always!

CHAPTER 3

BECOMING A PEOPLE PROFESSIONAL

"Render more and better service
than is expected of you if you wish to achieve success.
People who render the greatest service
also uncover the greatest opportunities."

NAPOLEON HILL

A professional cannot provide services or perform skills if clients, customers, or patients are not confident and if they do not have a high level of trust. Building a relationship of trust and confidence is perhaps the most critical step toward a person's acceptance of you and of the treatment you are providing. Without trust and confidence, you will never encourage a person to say yes to your recommendations.

Becoming an excellent provider of customer service (which is what being a "people professional" is all about) makes the difference. Remember that a patient might not know that you spent extra time and put extra effort into those margins. But I can almost guarantee you that they will know how they were treated on the telephone, how they were greeted, how comfortable they felt in your chair, how well you listened to their concerns, how respected they felt in your office, how you filed their insurance, handled their financial situation: *everything but the dentistry*.

Please do not misunderstand what I am saying. The dentistry has to be absolutely wonderful. However, people will base their decision on whether or not to proceed with treatment, whether or not to stay with you for the long haul, and whether or not they will refer other people to you based on the dentistry—and a whole lot more. That's where the people skills come in. That's where being a "people professional" makes the difference.

What can you do in the dental environment to encourage and nurture that valuable, yet elusive, quality of trust? What can you do to enhance your "people skills" so that you become a people professional?

Ten Ways to Become a People Professional

1. Define your ultimate mission, or your purpose. Write a statement of mission that becomes the foundation of all that you offer. Then commit to accepting no less of yourself or the members of your team. Define in your mission statement the type of care you will be offering and how that care will be focused on the service of your patients. As Stephen Covey says in his marvelous book, *Principle-Centered Leadership,* "A personal mission statement based on correct principles becomes a personal constitution, the basis for making major life-directing decisions, the basis for making decisions in the midst of the circumstances and emotions that affect our lives. It empowers individuals with timeless strength in the midst of change."

   Make sure that you offer only the best possible care—without compromise. Let this commitment to quality shine through in everything that you do, from your facility, your team members, your written correspondence, your treatment, and all follow-up. Offering only the best will make a statement to the patient, such as:

> *"We think you are special and way above average. Therefore, our goal is to provide you with special care that is way above average. You are worth the very best."*

2. Make sure that you are providing the type of care that patients want and expect. Find out how you are doing by asking questions or by asking patients to complete performance questionnaires (see patient questionnaires in Chapter 10). Do not become defensive by the responses. Use this vital information to improve your performance. Make sure that you are serving the needs of your patient base.

   Determining what your patients want can also be determined by very carefully analyzing their behavior.

   For example: What type of care do they readily accept? Do you need to offer more of that type of treatment? Do they keep their appointments? Do they value your time? Or does this need some work? Do they pay for their treatment—happily? Or do you need to work on financial arrangements? Are other payment options necessary to make the dentistry financially comfortable for your patients and for you?

   Asking these types of questions and objectively discussing them may give you clues as to what the patients are requesting. If those requests and the answers that follow fit into your statement of mission, then you have valuable information for practice enhancement.

3. Apply etiquette in all areas and in all relationships in the practice.

   a. Study and practice excellent telephone etiquette (see Chapter 9, Communication By Telephone). The person's first impression of the practice usually occurs during that initial telephone call. Make sure that everyone is answering the telephone in the same manner every time. Enthusiasm, warmth, and concern need to be combined on each phone conversation.

Careful messages need to be taken in a message book that has carbon copies. Post messages so that the responsible party can return telephone calls from patients promptly. You always want patients to know that their calls are important and getting in touch with them to answer questions is not only okay, it is desired. Make them feel as special as they really are!

b. Stand and greet each patient as they come into the practice. Greet them by name as you welcome them into your dental home. Sincerely inquire about their well-being and about their family. Know something personal about all of your patients. Record this personal information in your computer or on a special form in your charts. Become aware of this personal information during your morning meeting. Then when they arrive, make them feel at home by mentioning or inquiring about something personal. Make sure that what you mention is positive. You will want to start the appointment on a positive note. Treatment will proceed much better with a positive mind-set in place.

c. Introduce yourselves to all new patients. No matter what your position in the practice, introduce yourself and let the patient know about your position.

   For example: "Mrs. Jones, I'm Cathy. I'm Dr. Jameson's clinical assistant and I will be working with you today. We're glad that you've chosen our practice for your dental home, and we will do our best to make your time with us pleasant. You may come with me now."

   This approach is positive, encouraging, straightforward, informative.

d. Doctors, introduce yourselves, or make sure that the team member who is with you during your first exposure to the patient introduces you. I recommend a handshake. Why? You will soon be entering this person's mouth. Remember, the oral cavity is an intimate zone of the body. Therefore, it is critical that you establish a physical contact before providing your oral exam. The handshake is proper etiquette. But it becomes even more essential as you understand the body language of dental care.

4. Provide a full range of services: professional, financial, and emotional.

   As a people professional you are concerned with the total person. You understand the connection between oral health and total wellness. You also understand how a person's self-image and self-confidence are enhanced by or degraded by the smile. A person's comfort with their smile affects their sense of well-being.

   Because of the connection of the oral cavity and the smile to the entire person, you need to offer a full range of services or be able to make those services available through careful and confident referrals. In addition, present a full range of financial services so that the financial needs of the vast majority of your patients can be met. Establish a goal not to have patients walk out your door because they "can't afford it." There are financial options available today that meet the needs of most people. At the same time, these options are a benefit to the practice. You get paid, but the patient is not financially burdened (see Appendix A, Controlling the Financial Arrangements in Your Practice).

   When you carefully determine a person's (1) perceived need and their (2) clinical need through tedious diagnosis and treatment planning, and when you make the (3) financing of the dentistry comfortable for the patients and for you, you will be meeting the emotional needs of your patients. The three go hand in hand.

5. Spend time during your examination and consultation to educate patients about the recommended treatment. Be sure that they have "bought into" the treatment plan before you proceed. Involve the patient with the decision-making process so that they are clear about the treatment and so that they will feel essential to the co-diagnosis process.

   The initial examination and the consultation appointment are crucial. The time you spend to establish a relationship with the patient, the effort you put forth to determine the patient's perceived need, and the care with which you deliver your presentation will make or break the relationship and will certainly have an impact on whether or not a patient accepts treatment.

This careful and sincere attention to detail gives the patient a sense of value, worth, and respect. The respect you share with a patient will come back to you multifold. The respect with which you treat your patients will come back to you in the form of respect that goes far beyond the title and degree. It will be a respect that only comes when two people share common concerns and feelings, and when they work together to create a solution.

In today's health-care world, patients are looking for that respect. They are looking for the person who will give them the time to listen and to explain so that questions are not left unanswered. The listening and speaking skills that you will learn in this book will become a part of your patient relationships and will assist you on your journey of becoming a people professional.

6. Insure that continuity of information is evident. All team members must be on the same wave length and must be sending the same messages. An understanding of the services rendered must be an ongoing part of the education of the entire team. Appropriate questions need to be asked consistently. Proper verbal skills need to be designed, practiced, and implemented so that everyone is saying the same thing with conviction.

   This consistency must be evident whether a patient is in the clinical area or the business area. The business team needs to give backup support for the clinical team and vice versa.

   For example: When a patient asks the business administrator, "Do I really need this?" the business administrator needs to give affirmative encouragement and relevant information. In order to do so, she needs to know what is going on, why it is going on, what the benefits to the patient will be, and what the risks might be should the patient decide not to proceed. This third-party reinforcement can be very important in the patient's decision-making process. Remember: People look toward all members of the team as the professionals—and as such, they want and need your advice, your answers, and your encouragement. They want to know, "Am I making the right decision?" They are asking for help. Give it. Consistently.

If there are any areas of discussion or treatment that can (legally) be performed by an auxiliary person, then provide the necessary instruction and education and practice to raise their confidence level and the confidence level of the doctor. When patients receive consistent, quality care from all members of the team—not just the doctor—they feel totally confident and will develop an even greater bond to the practice.

7. Make sure that the completion of treatment obtains excellent results. Let the patient become overtly aware of the success of the treatment both verbally and visually. Take before and after photographs, or use an intra-oral camera with before and after images to validate the excellent results to yourself and to your patients.

   Evaluate your work with the before and after photography or with the intra-oral camera. In addition, radiovisiography gives you unparalleled diagnostic and evaluative capability. Use these new high-tech pieces of equipment to diagnose excellently and comprehensively, to evaluate your own work, and to give the patient visual evidence of the excellence of the treatment. Also, seeing the results of treatment will offset buyer's remorse if patients are wondering why they invested "all this money" in their dental care.

   Express your satisfaction with the results of treatment (backup support by the entire team is valuable here). Talk about the treatment and the positive end result. The patients will be reinforced to know that you are proud of the results.

8. Be willing to stand behind all treatment. Assure the patients that you are always there to answer questions and to support the treatment that you have provided. This commitment to the results will go a long way to build the person's confidence in you. They will know that if you are willing to stand behind your treatment that you will do the very best job—the first time.

In addition, if something does go wrong or if questions do arise, they will not feel that they are left out in left field to "root, hog, or die."

9. Stay in contact with your patient family in a positive way on a regular basis. Be sure that they hear from you through verbal and written contact (see Fig. 10–4 on page 139, Patient education newsletters). When people think of the dentist, you want them to think of you. And when they think of you, have them think of you in a positive way, not as the guy who hurts them or the guy who they only hear from when there is money due or the last person in the world they would want to see.

   Change the historical view of the dentist. Become proactive in your own way to change the patient's perception of who you are and what you do. Inform your patient family about the new and exciting things that are happening in dentistry today. Be excited about the possibilities. The more you are excited about new advances and opportunities, the more excited your patient family will become. Don't count on the outside world to educate your patients about dental advancements. You must accept the role of educator and assume the responsibility for this process.

   In surveys by the American Dental Association, four major factors emerge as deterrents to people seeking and accepting dental care. Those four factors are:

   1. No perceived need (lack of dental education)
   2. Fear of the cost
   3. Fear of the dentistry itself
   4. Time and convenience

   If lack of dental education is the dominant reason why people either don't go to the dentist in the first place or don't say yes to your recommendations, then it makes good sense to become great educators of dentistry. Become teachers. Learn the different personality styles so that you can individualize your instruction. Gear your presentation to the needs of each unique individual. Use excellent presentation skills (see Chapter

10, Making an Effective Case Presentation) and have fun with the education process.

Education is affected by all four of the communication skills: reading, writing, speaking, and listening. In fact, here is a breakdown of learning as it relates to the use of each of those four skills:

1. Writing: 9%
2. Reading: 16%
3. Speaking: 35%
4. Listening: 40%

All are critical to the educational process. However, most people will agree that listening may be the most essential communication skill of all. You will need all four communication skills to properly and effectively educate, and you will need to incorporate multiple types of educational tools and processes that encourage the learning process.

Taking time and applying excellent care to your educational processes will lift you up in the eyes of your patients and will enhance your patient care. Great people skills!

10. Provide the kind of care set forth in your statement of mission *always!* Be consistent in all that you do. Have all of your management systems in alignment and make sure that they are functioning well. Study and practice excellent communication skills so that you can accurately determine the needs of your patients and so that you can effectively deliver your messages in a clear and understandable manner.

    Give your patients what they want, need, and expect. Then give a little more. Always give more than is expected. It is the little extras that set you apart. Together, as a team, write out a scenario of what you think would be the ideal dental appointment for a patient. Make sure the scenario reflects what your patients have told you they want. Then design a plan of action as to how to make that happen every time. You want people walking out of your door thinking that they have never had such a positive dental experience in their life!

Once you have designed the plan, put the plan into action every time for every patient. Consistency is the key. People must not wonder what is going to happen in the office, they must be confident and sure of what will happen. Make sure that the "happening" is *fabulous!*

IN SUMMARY

Becoming a devoted and excellent people professional is fun! Your patients will confidently come to you, will look forward to their time with you (instead of dreading to come to the dentist), will be pleasant (most of the time), and will appreciate what you are doing for them. You will receive many more thank-yous when you provide care that is above and beyond the expected.

Do what is expected—and a little more.

Remember that valuable management principle. It is the basis of people professionalism. It is, indeed, "better to give than to receive." When you are in the business of giving your patients service based on quality, sincerity, and going the extra mile, you and your patients will feel good about your dental team. The rewards for being people professionals are personal and professional fulfillment as well as financial fulfillment.

CHAPTER 4

UNDERSTANDING PERSONALITY DIFFERENCES

"Temperament is the combination of traits we were born with; character is our 'civilized' temperament; and personality is the 'face' we show to others."

DR. TIM LAHAYE

FLEXIBILITY
VERSATILITY
ADJUSTABILITY

In order to function effectively with the majority of people, you need to learn and to apply these three functions: flexibility, versatility, and adjustability. Being flexible, versatile, and adjustable makes it possible for you to not only appreciate the differences among people but also to deal successfully with the uniqueness of individuals. You are not going to change people, nor would you want to do so. Your challenge is to adapt to individual differences so that you get along with and have successful relationships with individual teammates and clients.

On the one hand, differences can lead to conflict or to rejection. If this happens, the conflict usually stems from (1) misunderstanding of or lack of appreciation for variances in personality or (2) an incompatibility of goals.

On the other hand, understanding the differences in personality, appreciating those differences, and knowing how to relate better to each individual can lead to better problem solving, clearer and more enjoyable relationships, and greater case acceptance.

PERSONALITY DIFFERENCES

Misunderstanding of or a lack of appreciation for personality differences can be overcome if you study and accept the variances between and among people. Over 2,000 years ago, Hippocrates determined that there are four basic temperaments among men and women. The four basic personality temperaments or personality styles, according to Hippocrates, are:

- Choleric, the driving personality that wants results and control
- Sanguine, the enthusiastic personality that wants attention and positive strokes
- Melancholy, the steady personality that wants structure and organization
- Phlegmatic, the congenial personality that wants compatibility and harmony

Identifying these temperaments helps to clarify that each of your patients or teammates is unique and that treating each of them in exactly the same manner or presenting your recommendations in the same manner is inappropriate. As a dental professional your responsibility is to teach people about:

1. The needs in their oral cavity
2. What they need to do and why
3. What you can do to meet those needs
4. The financial responsibility and the benefits of making the investment

In order to meet these objectives and in order to achieve a high learning curve with your patient-students, excellent teaching must take place. One of the foundational principles of education is that you can't teach everyone in exactly the same manner. Different people learn in different ways. Thus individualizing your instruction is critical.

The first step in being able to individualize is to be able to identify each of the personality types or temperaments and to have a basic understanding of each. Then you can adjust your presentation modality or your approach to each person: individualization can begin. The end result will be that more people will accept your treatment recommendations, fewer people will reject these recommendations, and less conflict and fewer difficulties will erupt.

CHARACTERISTICS OF THE FOUR PERSONALITY STYLES

Choleric: Desires authority and prestige, takes a logical approach, likes a decision to be their own.

To relate most effectively with the choleric, you need to do the following:

1. Be brief and to the point. They want direct answers.
2. Stick to the business at hand.
3. Outline the possibilities.
4. Stress the logic of their decision.
5. If time is an issue, give them a time frame, but stress the benefits and end results of the treatment plan.

Sanguine: Cares about appearance, desires social recognition, likes to be recognized for their abilities.

To relate most effectively with the sanguine, you need to do the following:

1. Provide a friendly environment.
2. Let them express their likes and dislikes. Listen to them.
3. Show before and after photos or testimonials of others who have had the same treatment.
4. Give them a written treatment plan. They want reassurance. An intra-oral camera would be great for this person.
5. They want to feel respected and to feel a part of the decision.
6. Stress positives of going ahead with treatment.

Melancholy: Likes the status quo, wants security of a situation, needs time to think about this, wants to be appreciated, likes to identify with others, feels most comfortable with a specific pattern of treatment.

In order to relate most effectively with the melancholy, you need to do the following:

1. Provide a sincere, caring, agreeable environment.
2. Show sincere interest in them as a person.
3. Get their opinion by asking open-ended questions.
4. Be patient. They may not be sure what they want.
5. Give them a chance to adjust.
6. Define specifically what is going to happen.
7. Give them assurance and support.
8. Emphasize how going ahead will minimize the risk of getting worse.

Phlegmatic: Likes security, no sudden changes, personal attention, little responsibility, exact definition of their responsibilities, if any. They like a controlled environment and reassurance.

In order to relate most effectively with the phlegmatic, you need to do the following:

1. Take time to prepare in advance for any presentation.
2. Give straight pros and cons.
3. Reassure. No surprises.
4. Give exact details and precise explanations.
5. Provide a step-by-step approach to reach the goal.
6. If disagreeing, disagree with facts, not the person.
7. Be prepared to explain, over and over again.

How to Adjust to the Personality Differences

Let's take a situation and apply the knowledge of personality differences to the situation. In this way, you can see how the adjustments will be of benefit to you.

For example: A patient comes in for an initial examination. You provide that examination, develop a treatment plan, and invite the patient back to your office for a consultation appointment. The four following scenarios reflect an understanding of the differences in the patients, with the goal being to relate more effectively with each. If you present your recommendations in exactly the same way to each of the four personality temperaments, three of them will probably go away! If you don't adjust so that you relate well to your patients, they will not understand nor will they see the benefits and the relevancy of your recommendations. View an understanding of personality differences as a forward step in your efforts to enhance your people skills. Remember what John Rockefeller said: 85% of your success will be your people skills, while 15% of your success will be your technical skills.

Patient 1: Ms. Choleric

1. Have the following ready for her arrival:
   - written treatment plan
   - clear, written financial information
   - visual aids
   - scheduling alternatives
   - be well prepared for the presentation

2. Opening statement:

   DOCTOR: Ms. Choleric, last week when you were here for your initial examination, I asked you to tell me what your goals were for your teeth, your mouth, and your smile. You were very clear about your goals. I listened well and took notes during our conversation. During the past week, I have reviewed all of the diagnostic data and have designed a treatment plan to help you reach the goals that you have defined for me.

3. Present your recommendations quickly. Give specific ramifications: how long it will take, number of appointments. Show her examples of other similar situations so that she can see the results you intend to achieve. (But don't show too many examples. Be brief with this!) Give the financial responsibilities, close, schedule the first appointment.

   Be sure to let her know that you will provide as much treatment as is possible at each appointment, because you understand the intensity of her schedule.

4. Ask her if she has any questions or if you have responded to her requests adequately. Make sure that she knows that you have heard her, you respect her right to make her own decisions, and that you are a vehicle for her to reach her goals.

Remember: Be brief, to the point, direct, focused. The choleric is

1. Goal oriented: let her know you are in the business of meeting her goals.

2. Likes to be in charge and running the show: make her think that she is in charge and that you are there to help.

3. Strong willed and decisive: let her be involved with the decision-making process. You will get a lot further.

Patient 2: Ms. Sanguine

1. Be ready for her arrival by doing the following:

   - Make sure that you are totally prepared for her consultation. Have her greeted by someone she knows, perhaps a treatment coordinator who was with her during her initial consultation. Spend a bit of time with her during the beginning of the appointment to make her feel extra special and quite comfortable.
   - Have the written treatment plan available along with appropriate visual aids. If you have an intra-oral camera, have it in the consultation room with her full face image on the screen.
   - Have testimonial letters available from satisfied patients who have received similar treatment.

2. Opening statement:

DOCTOR: Ms. Sanguine, how are you today? (Then be a good listener.) We're glad that you are here today. I want you to know that I listened very carefully last week as we discussed your particular situation and your particular concerns. We took careful notes so that I could design a treatment plan that would meet your specific needs.

I know that you are uncomfortable with your smile right now and want you to know that we will work with you until we achieve the smile that you want.

In fact, Ms. Sanguine, I want to show you some situations that are similar to yours, people who were not pleased with their smiles and who are happy now as a result of receiving the kind of treatment that I am going to recommend to you today.

This person had a situation similar to yours (show the before photographs or images). Her front teeth had a space between them, and they were somewhat discolored. Can you see the similarity?

Once we completed her treatment, she looked like this (show the after photographs or images). What do you think? (Wait for a response.) Ms. Sanguine, I feel confident telling you that once we complete your treatment, you will look similar to this. What do you think? (Again, wait for her response.)

3. Then go into a discussion on treatment recommendations. However, be sure not to get too technical or too detailed. This patient wants to know (1) what she will look like, (2) if she will be comfortable during treatment, (3) if people will like her new smile. She is not interested in how many millimeters you will be removing or what wonderful technique you will be using. In fact, if you get too technical with this patient, she will not be able to make a decision. She will walk out of your door and you will wonder, "What ever happened to good ol' Ms. Sanguine?"
4. When you are discussing financial arrangements, be sure to reinforce the excellent decision that she is making. Give her permission to spend money on herself. Let her know that she will be pleased and that the people in her life will be pleased with the results. Ask her if she has any questions. Give her a card with a personally written note or name on it. Ask her to call with any questions.
5. Send her home with a coffee mug or other gift (see Fig. 4–1). Thank her for her time and for her confidence in you. Be sure to write her a personal note of thanks. Once her treatment is completed write her another note of thanks and send a before and after photograph to her place of employment. If she wants to tuck it away, she can (see Fig. 4–2). Even if she chooses to do this, you will be reinforcing her excellent decision to receive treatment. However, more than likely this exuberant personality will want to show other people her results. She will talk about you, because that is her nature. She will become an ambassador for your practice.

Remember: Be friendly, congenial, show her attention. The sanguine is

1. Talkative: ask open-ended questions to get to the bottom of their perceived needs.

Figure 4–1 An effective way to thank your referral sources and to build relationships with patients is to send coffee mugs. Your message will be replete as long as the coffee pot is on.

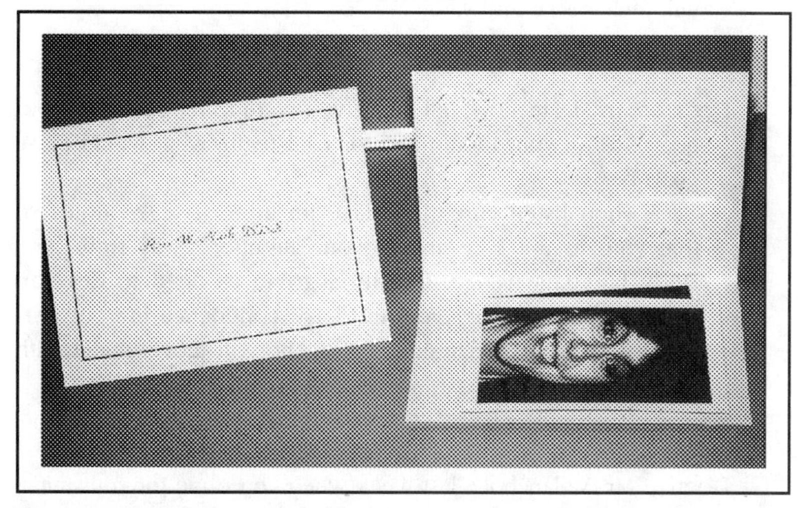

Figure 4–2 Reinforce the great end results of treatment with a personal note accompanied by a before and after photo (courtesy of Dr. Ross Nash).

2. Enthusiastic and expressive: show her wonderful before and after photographs to capture that enthusiasm.
3. Gregarious: appeal to her concern about appearance and about being socially acceptable.

Patient 3: Mr. Melancholy

1. Have the following ready for his arrival:

   - Upon confirming the appointment be sure to let him know that the doctor has reserved special time for him and that he will receive 100% of the doctor's attention. Have the written treatment plan ready for him and for the discussion. Greet him pleasantly, and make sure that you make note of something personal about him. Make notes during your initial examination about his personal situation and bring up something upon his return. He will appreciate your remembrance and will feel special and valued.

2. Opening statement of the initial examination:

   DOCTOR: Mr. Melancholy, tell me, if you were to tell me what your goals are for your mouth and for your teeth, what would those goals be?

   Then listen carefully. Have an auxiliary take notes so that you do not take any of your attention away from him. Body language is critical. He is going to need to develop excellent rapport with you and have confidence in your personal and professional skills before he will move ahead with treatment.

Opening statement of the consultation:

DOCTOR: Mr. Melancholy, last week when you came for your initial examination we gathered a great deal of data about your particular situation. You defined your goals for me at that appointment. During the past week, I have carefully evaluated that information and have designed a treatment plan that will let you accomplish your goals. Today, I would like to explain that treatment plan with you. I would like to discuss the possibilities, so that both of us are clear and confident that our work together will get you where you want to go. Would that be okay with you?

3. Then carefully discuss the following:
    - What he has now
    - What you need to do to help him reach his goals
    - What the benefits of treatment will be
    - What the problems might be if he does not proceed with treatment
    - The financial responsibilities
4. Ask for questions. Be careful about body language. Do not seem as if you are in a hurry. Give him time to think and to ask questions. He will need reassurance. Before and after photographs or images are, again, significant for validation with this patient.
5. When there are no further questions, ask if there is any reason why you should not proceed with treatment. Open the door for him to present any concerns, barriers, or objections. If you do not get these out in the open and if you do not have a chance to discuss these barriers, he will not proceed. Give him a copy of the treatment plan, and give him brochures specifically related to his situation. If he does not schedule the first appointment, ask permission to call within a week or so to answer any questions that he might have related to the treatment.
6. Be patient with this person. He will come around. It is just going to take a little time. He needs lots of reassurance.

Remember: Pay attention to detail, be serious, appeal to his intelligence. The melancholy is

1. Thoughtful: give him time. Don't push.
2. Organized: he likes lists, graphs, details: give him a written treatment plan with a written financial arrangement.
3. A perfectionist: make sure that you give him a sense of security about the treatment he will be receiving. Let him know that you will work with him until you get it right.

Patient 4: Mr. Phlegmatic

1. Have ready for the consultation:

   As with all personality styles—as with all patients—be exceptionally prepared for the consultation before the patient arrives. The phlegmatic doesn't like surprises, so you need to be ready to present all aspects of the treatment and be ready to answer many questions. He is not questioning you; he is just asking for help in making a decision. He needs reassurance.

2. Opening statement:

   DOCTOR: Mr. Phlegmatic, I've taken quite a bit of time over the last week to review the data we gathered last week. I wanted to be able to give you a clear picture of what you have now and what I am recommending as far as treatment. I'm going to get right to the point, and I'm going to tell you what I see as your major problem areas and what we need to do to restore your mouth to health again.

3. Then proceed with your explanation in detail much as you did with Mr. Melancholy. Describe the following in detail:

   - What he has now
   - What you need to do to help him restore his mouth to health again
   - What the benefits of treatment will be
   - What the problems might be if he does not proceed with treatment
   - The financial responsibilities

In your presentation to Mr. Phlegmatic, be sure to give the pros and the cons of treatment. He wants to know all the details. He wants and needs reassurance and wants no surprises, clinical or financial. Define and present your treatment recommendations in a step-by-step manner. Using a written treatment plan, determine what you need to do, how many appointments, the sequence of them, and how long each appointment will be. Then detail the fee for the total treatment and how much he will be responsible for at each appointment. Write the agreement. Give him a copy, and you keep a copy for your record.

Remember: Be gentle: the phlegmatic is low-key and easygoing. If you are of the high-strung nature—soften. Slow your speech. Be an encourager. The phlegmatic may seem uninterested, but keep encouraging. He may be shy and is probably indecisive. You will need to give strong support for your proposal and be assertive in asking him for a commitment to move ahead. That's okay. He wants you to make the decision for him and then be prepared to support him.

Delightful Differences

Study the different personality styles. Know that one is not better than another. There are, simply, differences. Understanding those differences gives the opportunity to relate better to each unique individual. Wouldn't you agree that relating better to a patient will give them a greater sense of security and trust with you?

In addition, understanding the differences and gaining the ability to be flexible will give you a much broader base from which to work. You will not change anyone. You cannot change a person's personality—nor would you want to. You, however, can adapt your *own* behavior and get along much better, present more effectively, and establish much stronger relationships. When people feel understood, they feel valued. When people feel valued, they will be more likely to make a commitment to you. Commitment means long-term relationships, working together to find solutions, trusting each other, and feeling confident with decisions made in harmony.

Understanding personality differences is a major step toward gaining greater acceptance of your treatment recommendations. This understanding is foundational to great communication.

CHAPTER 5

LISTEN YOUR WAY TO SUCCESS: REFINING THE ART OF LISTENING

"One friend, one person who is truly understanding, who takes the trouble to listen to us as we consider our problem, can change our whole outlook on the world."

DR. ELTON MAYO

Listen your way to success? How? How can this simple tool be considered the number one management tool in business today? How can listening effectively make a positive difference in your dental practice?

Lee Iacocca says, "Listening can make the difference between a mediocre company and a good company." Mark McCormack, author of *What They Didn't Teach You At Harvard Business School,* feels that "in selling there is no greater asset." Sperry, one of America's major corporations, considers the skill of listening so important to the success of their business that they have devoted extensive time and funds to the development of courses for the instruction of listening. They have made these courses available for all levels of personnel. Sperry believes that the inability to listen leads to such business inefficiencies as:

1. Wasted time
2. Ineffective operation of the departments
3. Miscarried plans
4. Frustrated decisions in every phase of the business

Remember: There are four operations involved in communication through words: writing, reading, speaking, and listening. When asked which of these four operations is the most significant for effective communication to take place, most people will say—without hesitation—listening. And yet most people feel that listening is the least developed and the least well performed of the four communicative operations. Listening needs to be more than an activity of the ear in order to be effective; it needs to be an activity of the mind.

What Is Listening?

Listening, according to Webster, is (1) paying attention to sound, and (2) hearing with thoughtful attention.

Listening, according to Kevin Murphy, president of CDK Management and Consulting Associates, is

1. The accurate perception of what is being communicated,
2. A process in perpetual motion,
3. A two-way exchange in which both parties involved must always be receptive to the thoughts, ideas, and emotions of the other.

Mr. Murphy notes that "Listening is a natural process that goes against human nature!"

Effective listening stimulates the team to generate additional creativity, to solve problems, and to execute smoother systems. In addition, effective listening lets you become aware of the needs and desires of your patients so that you can go about the business of meeting those needs.

LISTENING LEADS TO GREAT TEAMWORK AND BETTER PATIENT RELATIONSHIPS

The *team* is the lifeblood and the heartbeat of the dental practice. If, as the leader of the team, the dentist realizes this fact, taps the incredible resources he/she has available within the team, and genuinely listens to each member of that team, success *will* be realized.

Pooling the vast resources available on a dental team does the following:

1. Helps the dentist zero in on multiple sources of information
2. Helps the dentist to be a better employer
3. Helps the dentist to be a more successful business person

Ultimately, and most importantly, the patients benefit from excellent listening skills, because individual and unique needs, fears, concerns, and wants are truly heard. Only when these needs are truly listened to will you have the opportunity to do something to satisfy those needs.

By listening to your patients you will learn how they feel about your service. You will learn what it is they want. Then you will be able to respond. That's what customer service is all about. Defining the needs and meeting them. The patient wins by having necessary or desired treatment rendered and you, the dental team, win with increased productivity.

WHAT GETS IN THE WAY OF EFFECTIVE LISTENING IN THE DENTAL OFFICE?

Some of the main deterrents to good listening may be:

1. Time pressure
2. Stress, not being able to relax

3. Mind-set, being rigid in thought processes
4. Talking too much, dominating the conversation as the authority
5. Thinking what to say in response instead of listening
6. Lack of interest
7. Ego that says, "I know the answer or I know what to do, so I don't need to listen to you."

Before you can learn how to listen effectively, you must develop an understanding of the attitudes necessary for this type of listening to take place. Kevin Murphy in the book *Effective Listening: Hearing What People Say and Making It Work for You* says, "A mind is like a parachute—it only works when it's open!" An open mind is necessary for the skill of listening to be effective.

WHAT ATTITUDES ARE REQUIRED FOR SUCCESSFUL LISTENING TO TAKE PLACE?

1. You must want to hear what the other person is saying.

This takes time! If you don't have the time, be respectful and say so.

For example:

DOCTOR (TO ASSISTANT): Mary, I understand that you have a concern about the late hours we've been keeping! I want very much to hear what you have to say, the suggestions you have, and so on, but this is not a good time for such an important discussion. Could we get together tomorrow for lunch and concentrate on this issue—uninterrupted?

2. You must sincerely want to help the other person with the problem.

If you don't want to help, wait until you do!

For example:

DOCTOR (TO ASSISTANT): Sherry, I realize that the summer is a difficult time for you because your kids are home from school.

However, right now changing schedules is not a possibility, so let's table this discussion until next month.

Note: If you do tell a team member that you will get together later for the continuation of a discussion, *do so!* It will be debilitating to your relationship to say you will deal with something and then let it slip away. Respect will deteriorate.

3. You must be able to accept the other person's true feelings.

Other people will have feelings different from yours. Sometimes these feelings may be different from what you think they should be! Learning to accept these differences and not letting that difference affect your relationship takes time and effort and is not easy.

For example:

MS. PATIENT: I hate the dentist! I only come when I *have* to!

ASSISTANT: I understand that you have some apprehension about the treatment you are going to receive, Ms. Patient, but I want you to know that Dr. Best is great! He is committed to gentle, caring dentistry. We are going to help you understand the treatment, so that you will not feel so apprehensive about your time spent with us.

This patient felt differently about dentistry than the team member. However, this difference in feeling did not change the assistant's commitment to educating the patient about good dental care, nor did this difference in feeling negatively affect her attitude about the patient.

4. You must trust that the other person has the ability to handle his/her feelings and can deal with—or work through—the given situation.

When a person comes to you and expresses a concern or a problem, they aren't asking for your advice, they are asking for your attention and for your care. Most people do a sensational job of closing the doors to effective communication. They do so by throwing up barriers to communication.

Barriers to communication prevent a person from giving you further information—and information is exactly what you want. It lets you identify motivators, problems, concerns, or emotions.

Barriers to good listening are numerous. A few are listed below:

1. Giving advice
2. Offering a solution
3. Passing out orders or directives
4. Hitting the person with an ultimatum
5. Preaching and teaching
6. Being critical
7. Flattering
8. Putting the other person down for their thoughts
9. Patronizing
10. Making fun of the person or making a joke of their problem

By practicing excellent listening, you allow another person to deal with their own issues. You allow them to clarify some of their own questions. No one can solve a problem for a person except that person. Listening leads the way to defining the problem so that successful problem solving can take place.

Don't take away from a person's ability to deal with their own issues. Be an enhancer. Listen. Listen without judgment. Allow a person to have separate feelings from you and allow that person to deal with their own emotionality.

Oftentimes, listening to a person and letting them get something out on the table is all that is necessary. They are probably not looking for a solution: they are looking for a sounding board. Know that if people want your opinion, they will ask for it.

5. You must know that feelings are often transitory.

Accept that changing feelings are part of human nature.

For example: A member of your team may be suffering from burnout. They may say they want to quit, that they aren't happy or satisfied any longer.

Be patient. Listen. Respect the person enough to allow them to express their concerns. Determine if the problem can be solved or if it is too late for a solution. If there is a willingness to work on problem solving, develop a plan of resolution—together. The feelings may be transitory. The team member is too valuable to release because of this transitory emotional and physical state. Don't make judgments about this person based on your automatic reaction. Take the time to truly define the problem, design a plan for the resolution of the problem, and implement the solution. Many emotions are transitory. Accept the humanity of this.

6. You must be able to actually listen without becoming self-stimulated or defensive.

Allow for a "separateness." The other person is unique from you and responds in his/her own way. Respect this separateness.

For example: Oftentimes, when we hear what another person says to us, we become defensive and thus close the door on good communication. A more effective way to truly listen to another is to reflect back to the person that which we think we are hearing in order to get to the center of the message, what the sender really means.

Saying nothing at all *does* communicate acceptance, if you are truly listening attentively. Silence or *passive listening* is a nonverbal message that, when used effectively, can make a person feel genuinely accepted.

A CLOSER LOOK AT LISTENING SKILLS

Dr. Thomas Gordon, in his outstanding work on communication skills teaches that listening skills are the skills necessary to help another person when that person has a problem. Listening has three specific modalities:

1. Nonverbal attending skills (body language)
2. Passive listening
3. Active listening

NONVERBAL ATTENDING SKILLS

Research has shown that approximately 60% of the perception of a message, whether it's being sent or received, depends on body language. About 30% of the perception of the message is sent or received by the tone of the voice. Only 10% depends on the words themselves! If this is true, then one must pay close attention to and carefully plan the physical messages being sent and being received. You can make or break a conversation or a presentation by your body language. In fact, even if all of the words are great, the message can come across incorrectly as a result of body language and tone of voice.

Positive body language postures that you can use to express an attentive, listening posture are:

1. Establishing and maintaining eye contact
2. Positioning yourself on an equal level
3. Staying face to face
4. Touching firmly, but gently
5. Nodding or shaking your head
6. Facial expressions
7. Reflecting a person's body movements
8. Taking an open stance with your arms and legs, not crossing them or closing yourself off
9. Nodding or shaking the head

In being receptive to a person's comfort or understanding, there are numerous body language expressions that you can observe from the patient:

1. Arms and legs in a crossed stance
2. Clutching the arms of the chair
3. Frowning and other facial expressions
4. Voice inflection, tone, volume
5. Hand/body gestures

6. Posture
7. Touching behavior
8. Physical distance
9. Skin responses such as blushing or pallor

Do your patients send messages to you with body language?

The answer to this question is a resounding yes. Truly actions speak louder than words. You need to be aware of cues being expressed by your patients so that you can respond to their message.

Body language cues are sent by clients/patients that can tell you a great deal about their attitude toward the presentation you are making.

The following are body language cues to watch for, what those cues might suggest, and how you can effectively respond to those cues.

1. Cue: The patient's hands are open and relaxed with palms turned upward.

   Indication: Positive response, open to your recommendations, ready to work with you.

   Your response: Move on, mirror their actions.

2. Cue: No eye contact.

   Indication: Not comfortable with you or with something you've just said. May be intimidated by or insecure with you.

   Your response: Repeat your last comment or statement. Tell them that you want to take care of him/her. Try to relax the patients, smile softly. Gently but firmly touch his/her shoulder and ask if what you have just said bothers them.

3. Cue: Patient leans toward you or moves a little closer.

   Indication: Trust is being established. He/she is beginning to accept what you are saying.

   Your response: Ask questions to see if he/she is beginning to accept your recommendations (i.e., "Chewing better is what you want, isn't it?").

4. Cue: Patient is looking directly at you and is giving you his/her full attention. Touches chin or side of face.

   Indication: Very receptive, giving thoughtful consideration.

   Your response: Continue.

5. Cue: Takes glasses off and wipes the lenses.

   Indication: Thinking it over.

   Your response: Slow down. You are moving too fast. Reiterate and clarify your last point. Recap the points you've already agreed upon.

6. Cue: Eyebrows raise; forehead wrinkles; mouth falls open.

   Indication: Shock, surprise.

   Your response: Ask questions to find out if the surprise is positive or negative.

   For example:

   DOCTOR: From the look on your face, I sense that you are surprised. Are you?

   PATIENT: Yes. I'm very surprised!

   DOCTOR: Surprised in what way and about what issue?

   If the response is positive, go on! Encourage even more excitement. If the response is negative, calm and reassure him/her.

7. Cue: Pulls on ear while you are talking.

   Indication: Wants to break in to ask a question or make a point.

   Your response: Pause. Ask if he/she has a comment or question.

8. Cue: Hand covers mouth.

   Indication: Insecure, self-doubt, self-conscious.

   Your response: Reassure. Go back and clarify points you've already made.

9. Cue: Squirming in chair.

   Indication: Feeling pressured or uncomfortable.

   Your response: Reflect back to the person what you think they might be feeling. Clarify if your perception is accurate or not. Find out what's making them uncomfortable.

10. Cue: Fidgeting with an article such as a bracelet or watch or pencil. Eyes lowered as you are talking.

    Indication: Insecure, lack of confidence.

    Your response: Validate the person, give positive reinforcement about something. Explain your recommendations again or in more detail (i.e., "Ms. Patient, let me go over my recommendations with you again. That might answer questions you have at this time. Would you like that?").

Don't underestimate the power and significance of these body language messages or sensory perception cues. The acceptance of your recommendations will improve in direct proportion to your ability to relate to and with your patients. Don't miss an opportunity to listen accurately to their body language.

The giving and receiving of messages can be an asset to the total dental experience. You, as a good listener, want to perceive all possible messages. You will then be better able to plan your presentation of recommendations and the treatment itself. And the more a patient feels listened to, the more comfortable, relaxed, open, receptive, and less difficult he/she will become.

GO ON!

The secret to getting along well with other people is to determine their needs and then to be willing to help them satisfy those needs. How do you find out what those needs might be? Ask open-ended questions, questions that cannot be answered with a yes or a no. Then *listen* to pick up information that might (1) give you a clue as to what motivates that person, (2) what they want, and (3) what the possible barriers to acceptance or a solution might be.

Open-ended questions or statements that encourage a person to continue providing information might be:

1. "Tell me, how we can help you."
2. "Tell me about it."
3. "Give me some more information."
4. "Go on."
5. "I'd like to know how you feel about that."

There are ways of listening that encourage a person to continue, to go on, to get it all out. Ask a question and then respond with a non-stimulating response. This type of receptive listening means that you are attentive, you are hearing what the other person is saying, but you are careful not to interfere with their commentary. Dr. Gordon calls these *acknowledgments*. A few of these encouraging responses might be:

1. "I see."
2. "Really."
3. "Uh, huh."
4. "Hmmm."
5. "I agree."
6. "Continue."
7. "Tell me more."

Pay attention to the body language with appropriate attending skills, listen caringly and carefully, and let the other person continue without interruption. Don't try to fill all silent moments. Silence is a communication skill. Careful pauses in the conversation give people time to organize their thoughts, to gain the confidence to go on, and to express their deeper feelings. Silence or passive listening can be a form of very powerful communication.

If you want to be receptive in your listening and encourage the person to "go on," then this type of "passive listening" will do just that. According to Dr. Gordon, "Silence or passive listening is a potent tool for getting people to talk about what's bothering them, and talking to someone who is willing to listen may be just the encouragement a person needs to keep going."

The most challenging part of passive listening is asking the question and then waiting for the person to respond without interrupting! When you ask that question, *be quiet!* Wait for an answer, even if a few moments of silence occur.

ACTIVE LISTENING

When you have opened the lines of communication, and have encouraged a person to "go on," the next step is to listen actively. The single most effective listening skill you can use to calm an irate person, to defuse anger, to handle a difficult person, or on the other hand, to enhance a good relationship, is to listen actively or reflectively. This kind of listening requires effort and discipline.

Active listening according to Dr. Gordon is "feeding back to the person what you think you have heard them saying to make sure that you have heard them correctly." A simple repetition or paraphrasing is not sufficient. The listener should demonstrate, in his/her own words, an adequate understanding of the content, intent, and emotion of the speaker's remarks.

This type of listening gives you the opportunity to clarify, to make sure that you understand what the person is trying to say. It has a definite calming effect by providing feedback and by maintaining a supportive atmosphere.

For example:

MS. PATIENT: At my last dentist's office, I could just pay him after my insurance had paid. I can't believe that you are suggesting that I must pay my portion the day I receive this treatment. This is infuriating!

BUSINESS MANAGER: Ms. Patient, you seem upset because our financial policy is different from the one you are used to.

MS. PATIENT: I am! I don't see any reason to pay anything until I know exactly what my insurance is going to cover.

BUSINESS MANAGER: I understand your confusion over the difference, Ms. Patient. However, although we file your insurance as a service, you are responsible for the portion that insurance does not cover. Therefore, we ask that you take care of your estimated portion the day of the service. We want you to be aware of this before treatment is rendered so that there will be no surprises for you following your treatment.

MS. PATIENT: Well, I can appreciate that! So, next week when I come in for my appointment I need to pay $______, is that right?

BUSINESS MANAGER: Yes, that is correct. We will file your insurance that day, also. Then if there is any difference after your insurance pays, we will immediately notify you so that you can take care of that difference.

By using active listening in the beginning of this conversation, the business manager was able to figure out the problem. She didn't antagonize the patient, which would have led to more anger, a possible scene, and the chance of losing the patient.

Active listening is used when someone comes to you with a problem or a concern. Active listening is the restating in your own words what you understand the other person to be saying. In actively listening to someone, you need to give careful attention to both the content of the message and to the feeling that is being transmitted.

The following examples will display a message sent by a person *who owns a problem*. The responses that follow are examples of (1) a response that is typical—and yet acts as a door closer, and (2) a response that will illustrate active listening—a true door opener to communication.

Situation 1

TEAM MEMBER: Why does Sarah have to make so many mistakes?

OFFICE MANAGER 1: She is having a lot of problems at home. Give her a break!

OFFICE MANAGER 2: You're feeling frustrated by Sarah's performance these days? (active listening)

Situation 2

ASSISTANT: I'm not sure I'm ready to assist you with that procedure yet.

DOCTOR 1: You can do it if you just try!

DOCTOR 2: You're feeling uncomfortable because you're unsure of how to perform this procedure? (active listening)

Situation 3

TEAM MEMBER: I won't ever say anything in a staff meeting again!

OTHER TEAM MEMBER 1: Well, that wasn't a very good idea you threw out!

OTHER TEAM MEMBER 2: You feel hurt because you are not sure your ideas are accepted. (active listening)

In the previous examples, the first response would probably *not* encourage further discussion nor would the response open the door to positive communication. If there was a problem to be solved, the first response did nothing to further the seeking of a solution. However, the second response— active listening—indicates a true acceptance of the person's feelings and an understanding of the content of the message. Now they can go on to the defining of the problem and to the solving of the problem.

When you learn to integrate active listening into your repertoire of communication skills, you truly involve the mind in a dynamic process, rather than using only the ears in a physical process. You reach out to the sender of the message with your own message of caring and acceptance.

If you use only your ears to hear the words, but do not use your mind to understand what is *really* being said and felt, then you do nothing to advance communication. The result is "failure to communicate!"

From this type of feedback the sender gets tangible evidence of how the receiver deciphered the message. The sender can either confirm the accuracy of the message ("Yes, that is just what I meant.") or can deny the accuracy ("No, what I meant was—"). This kind of continual feedback allows you to be absolutely sure that you understand what is being said. This also provides a sense of empathy and acceptance for the one delivering the message.

One of the main purposes of active listening is to keep misunderstandings to a minimum. When you are involved in a listening situation, make sure that you do so with sincerity, understanding, acceptance, and caring. Listen to your teammates and to your patients—*open doors* to communication!

You are in the profession of dentistry, and no matter what the specific goals and objectives of your practice, each and every one of your days is an exchange of context and emotion—or *communication*—both verbal and nonverbal. If you choose to ignore or underplay this constant exchange of data, you will be missing an incredible opportunity to grow your practice by using numerous information sources, by solving everyday problems constructively and painlessly, and by drawing upon the talent that penetrates your practice every day!

Put the greatest single management tool you can possess into effect: *listening*. Put it to work *for* you! Other successful business leaders have discovered its value; so can you! You will be encouraging (1) the creative potential of your employees, (2) a strong, enthusiastic team, (3) patients to accept treatment and optimum dental health, and (4) the reaching of your own and your practice's potential.

Chapter 6

Getting Your Message Across in a Positive Way: Speaking Skills

"The meeting of two minds may consist in their understanding one another while still in disagreement or it may consist in their coming into agreement as a result of their understanding one another."

Mortimer Adler

Paul Harvey says, "It's not what you say, it's *how* you say it!" I totally agree. So many times we taint our message by interspersing words and phrases that conjure up a negative response or a negative thought.

As dental professionals, your hearts are in the right place. You want to help people. You want patients to understand that everything you do is geared toward better care. But sometimes you say things in a way that generates a negative response from the person to whom you are delivering the message. Improving your speaking skills can become an asset to your relationships and to your practice.

THE COMMUNICATION FLOW

Excellent communication takes place between two people when the message sent by the speaker is interpreted correctly by the listener. Obviously, both skills—listening and speaking—are critical for the communication flow to work.

Speaking is a part of the communication flow. Speaking so that others will listen to you and so that your message will be understood is a critical part of that communication flow. In each and every one of your dental days, all members of the team have the opportunity to move a relationship with a patient further along or to pull that relationship to a stop by the way they communicate with that patient.

In addition, how each team member speaks to the patient can make a difference in whether or not a person accepts treatment. The goal of each member of your team needs to be to listen well enough to determine the patient's perceived need or their own goals for their oral cavity and to speak professionally and compassionately so that people understand the benefits of your services and will choose to proceed with treatment.

The goals of good speaking are:

1. To help people *want* to listen to what you have to say
2. To deliver the message in the best possible manner
3. To check to see if you were heard (or interpreted) correctly

CRITICAL FACTORS IN THE COMMUNICATION FLOW

There are three specific factors that define whether or not a person will want to listen to the message you are sending. They are:

1. Is this of interest to me? Will this benefit me?
2. Who's delivering the message?
3. How is the message being delivered?

1. *Is this of interest to me?* People are motivated by "what's in this for me?" People want to know how something will be of benefit to them. Is the information useful, fulfilling, productive for them? If someone is *told* what's good for them, they probably won't buy into it. A person must see how a product or service applies to them and how it will benefit their particular situation. In other words, people want to know, "How does this affect me?"

In order to get people to the point where they are willing to listen to your proposal, you first have to determine what motivates them. The only way to determine a person's motivational "hot buttons" is to ask questions—and listen. Once you have determined what those hot buttons are, you are in a position to respond to their needs. You can't push anyone into making a decision, but you can lead them to making a decision by asking questions, determining motivators, and delivering your proposal as a direct response to their needs.

Make it interesting:

- Ask questions to open the lines of communication.
- Determine the motivational hot buttons.
- Respond to the person's needs in your presentation.
- Focus on how the recommended behavior will benefit the person.

2. *Who's delivering the message?* Does the person receiving the message trust the person sending the message? Your initial contact with a person is very critical. A person needs to know your intentions before they can *or will* trust you. People want to know, "Are you for me ?" "Will you help me?" "What will happen if I let you in?" "Will you take advantage of me?"

People decide if they trust you in two ways: emotionally and intellectually. Most decisions are made emotionally. Even though a person tries to make a decision based on fact, the process is filtered through the emotions.

A level of trust must be established before you can gain another's confidence. Don't underestimate the importance of spending time and giving attention to the building of rapport. The time spent on relationship building may be the best time you spend with a person. You will only go so far with a person until they are confident in your sincerity and in your true interest in them. The level and the depth of your relationship will be in direct proportion to that level of trust.

3. *How is the message being delivered?* How do you get your message across to the listener or receiver? Remember that 60% of the perception of a message is sent or received by body language, 30% through the tone of voice, and 10% through the words that are spoken. The way in which you deliver your message may well have more impact on the listener than what you are saying.

In addition, people learn and respond more effectively to visual messages than to auditory messages. Approximately 83% of a person's learning takes place visually. Therefore, supporting your information with visual aids will enhance your ability to clarify your message, to have a better impact, to make your point effectively. This allows the listener to be involved with the communication. Involvement is a key to learning and understanding.

There's the goal: to have communication experiences become a mutual involvement of all parties, to have interactions that encourage the giving and receiving of information and feelings. The result? Winning relationships. Strong relationships. Acceptance of treatment proposals.

Remember:

- Be direct with your message. Don't beat around the bush.
- Be specific. Get to the point.
- Be positive.
- Don't use put-down language or a patronizing tone of voice.
- Learn what motivates the person.
- Establish a relationship of trust.
- Present your information succinctly and visually.
- Involve the person throughout the conversation.

HOW TO SEND CLEAR MESSAGES

How you communicate with your patients can be a source of mutual respect and understanding or can be a source of frustration and discouragement for the both of you. Your knowledge and skill of communication is the bottom line of successful relationships.

Effective communication takes place (as illustrated in Fig. 6–1) when there is a clear sending of a message (speaking skills) and appropriate feedback to check the accuracy of the reception (listening skills). In other words, "Did you get my message?" and "Am I hearing you right?" It is definitely a two-way process.

A person sends a message either verbally or nonverbally—or both. The information that the speaker wants to convey to the receiver is transformed into outward behavior (verbal or nonverbal).

The message is received by the other person who then interprets

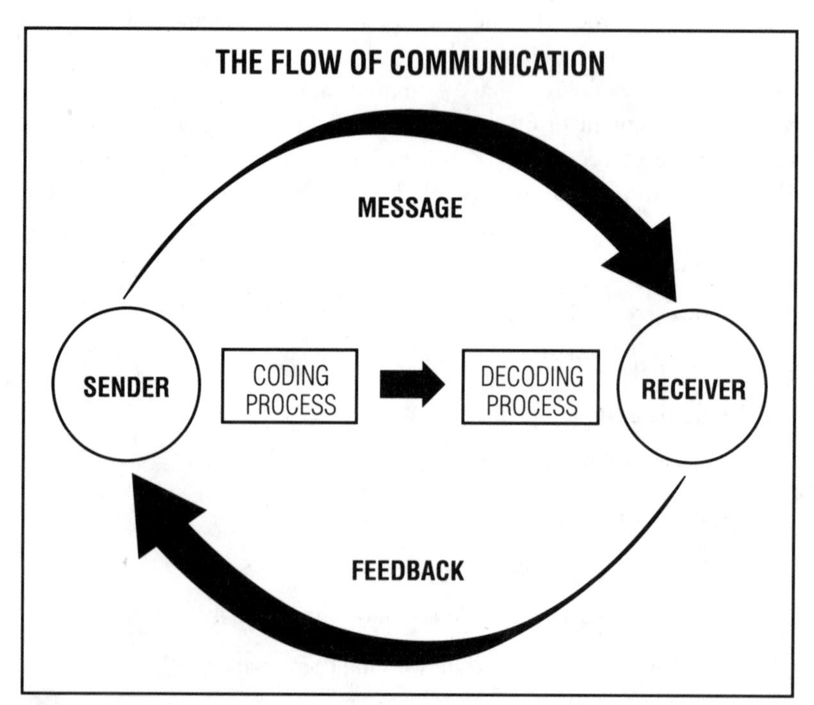

Figure 6–1 Communication is a two-way dynamic process.

the message based on prior experience and understanding. Feelings are produced in reaction to the interpretation. Judgments are made about the intention of the message.

Receiving or requesting feedback gives the sender of the message a chance to know if the message has been heard accurately or if it needs to be clarified. This feedback may be the best communicative skill you can ever acquire—the skill of determining if your message has been sent and received accurately.

For example:

DOCTOR: Mrs. Jones, during our initial examination you told me that you wanted to keep your teeth for a lifetime. That's great. We want the same thing for you. Based on my diagnosis and evaluation of your situation, I am going to recommend that we become involved with an active program of periodontal therapy that will let us get and keep your gum tissue healthy.

Without healthy gum tissue, we cannot restore your teeth to health again. We must have a solid foundation upon which to work. This solid foundation will secure the investment in time and money that you will be making, and it will serve as security for your health. Getting and keeping a healthy mouth is your major goal, isn't it? Then, if it's okay with you, I would like to explain what is involved in periodontal therapy and how the therapy will benefit you. (Explanation follows.)

Mrs. Jones, do you have any questions about the therapy I am recommending for you? (Pause for questions.) What aspects of the therapy do you think would be the most beneficial? (Pause.) What areas cause you the most concern? (Pause.)

(Doctor asks for feedback. If the patient is clear about the message he/she has sent, then the Doctor proceeds.)

7 STEPS OF EFFECTIVE SPEAKING

STEP 1: Actively listen

Use feedback. The only way you will know if you are getting your message across accurately is to ask for feedback. In order to make sure that your message was clear, ask questions like the following:

1. "Does that make sense?"
2. "Do you have anything to add to this?"
3. "How do you feel about that?"
4. "Did I say that clearly or do you need more information?"

STEP 2: Benefit statement

Make a statement that reflects the benefits of the information you have just shared.

> DOCTOR: Mrs. Jones, you seem to be clear about the recommendations I am making for you. I feel that this treatment will make it possible for us to save your teeth. This will prevent the need for more extensive treatment in the future.

STEP 3: Close your benefit statement with a question

Why do you need to close your benefit statement with a question? To accomplish the following: (1) to get the patient involved, (2) to see if you were clearly understood, and (3) to find out if questions or concerns still exist.

> DOCTOR: Mrs. Jones, do you have any questions about the treatment that I am recommending? (Pause.) If there are no further questions, is there any reason why we shouldn't schedule an appointment to begin your treatment?

STEP 4: Use a variety of teaching/communication methods

Because people learn in a variety of ways and because people have different behavioral styles, you may want to approach your communication efforts in multiple ways.

For example: If you are presenting a treatment plan, you may want to use before and after photographs to back up your verbal presentation. Or you may find that certain people want only the facts while others may want a great deal of detail (see Chapter 4, Understanding

Personality Differences). You must be perceptive to the different behavioral styles and respond appropriately.

STEP 5: Notice the body language

Pay attention to how the receivers of your message are responding. Are they making eye contact with you? Are they fidgeting? Do they seem preoccupied?

Remember to ask for feedback. See if you are coming across well. If the person does not seem to be in touch with you, gently acknowledge this by asking: (1) "You seem to be preoccupied. Is there something on your mind?", or (2) "This doesn't seem to be a good time for you. Would you like to postpone this conversation?"

There is no reason to proceed or to think that you are going to be heard if you do not have the other person's attention. The human mind can only think of one thing at a time. Take a few moments to get the person's attention, then proceed—or reschedule the conversation or consultation.

STEP 6: Use layman's language

You've heard this a million times. That is because it is so important! Dental professionals often think they are using layman's language when in fact, the patient has no idea what you just said.

The English language often has multiple meanings for one word. Some words have as many as 25 different meanings. Often when a dental person is explaining treatment, the person becomes lost and can never really get back on track. The patient may be embarrassed to tell you that they do not understand and so they just say, "Yes, I understand." Then they leave, and you wonder what happened to them. Know that just because people say they understand doesn't mean that they do.

Tape record yourselves in various situations: a consultation appointment, making a financial arrangement, delivering hygiene instructions, giving postoperative instructions. Then replay the data as a team. Try to determine which words might throw off a patient. Change those words or phrases to be more user friendly.

You might even consider playing your tapes for a nondental person. Let them tell you what makes no sense or what is confusing. This effort could prove to be extremely valuable to you.

STEP 7: Give positive reinforcement

Thank the person for his/her concern, attention, and participation. Again express the value of the information you are sharing. Reinforce the benefits of your recommendations. Repetition is essential to learning.

The goal of effective speaking skills is to open the doors of communication rather than to close them. You want your messages to come across in a way that people will listen and understand. Making sure that you are understood and accepted is no easy task. Just talking or lecturing will not get the results you want. Learn how to speak so that others will listen.

ADDRESSING CHALLENGING ISSUES

In Chapter 4 you learned how to listen so that you can be clear about what other people are telling you. You learned the skills of listening so that you can be more helpful and more effective in your relationships with team members and with patients when they have concerns or problems. In this chapter you have learned the speaking skills to deliver your message more clearly. However, you may now be asking yourself, "Great! I know what to do to help others if they have a problem. I know how to speak effectively in a presentation. But what do I do when *I* have a problem with someone or I need to let a person know that I have a problem? Are my speaking skills important in that type of situation?"

The answer is undeniably *yes!*

If someone's behavior is having a concrete, negative effect on you, on your performance, or on the practice, then you have not only the right but also the responsibility to address the issue. It's imperative that you address the behavior and not the person. Some people have a difficult time hearing about their own performance. They take "constructive criticism" personally and become offended by the confrontation. Therefore, learning to speak in a nonthreatening manner will move you closer to the goal of being able to confront constructively—not harmfully.

You need and deserve to have your needs met. In order to be successful in getting your needs met, certain speaking skills are necessary and beneficial. Developing excellent speaking skills will accomplish the following goals: (1) let you get your message across accurately, (2) have your needs understood and met, and (3) move your relationships to the next level, to gain a person's trust and confidence.

The specific communication skill to use when someone is causing you a problem is an "I" message as developed by Dr. Thomas Gordon. In his book *Leadership Effectiveness Training,* he suggests that when you *own* a problem (that is, when what someone is doing is having that concrete, negative effect on you, your performance, or the practice), that you need to use assertiveness skills. Assertiveness skills indicate that you will convey your needs and work toward getting those needs met. At the same time, you will do everything possible to help the other person get his/her needs met as well. Both parties will win.

Not confronting a person whose behavior is causing you a problem doesn't serve either party well. If people don't know that what they are doing is inappropriate, they can't change. If you do not confront difficult behavior, you run the risk of harming and even losing the relationship.

People become uncomfortable with confrontation because they may have received negative reactions in the past. If you have had less than great confrontational experiences, it may be because either you or the other party—or both—have not had the skills necessary to confront constructively and positively.

It takes a great deal of courage to confront another person about their behavior. There is risk involved. You have to decide whether or not the person's behavior is having a concrete negative effect on you or not. If it is, then the risk of confrontation becomes necessary. If the answer is no to that question, then let it go! That may be one of the most difficult things you ever do: let go of something that may be bugging you but is not truly a problem. In the one situation where you choose to risk confrontation, your goal will be resolution, and the skills you use will be critical to your success. In the other situation, where you choose to let it go because you have determined that this person's behavior is not really causing you a problem, great stress relief may result. Do you carry around burdens, worries, upsets, discomforts for no concrete reason? Do yourself a favor and let them go.

I-MESSAGES

I have referred to Dr. Thomas Gordon many times in this book because I am a certified instructor of Dr. Gordon's Effectiveness Training. Over the last 20 years, I have taught people the skills of effective communication. The skills have become a part of who I am and certainly a part of how I communicate. These skills have been an asset in my family life as well as my professional life. In learning to confront the challenges of everyday life, one of the most beneficial skills I know is the I-message.

In seeking solutions the I-message is not the problem solver, but it is an excellent way to get the problem defined or clarified. That is the first step in problem solving: to define the problem. An I-message is a clear, accurate code sent when someone is causing you a problem. It is a message made up of three specific parts:

1. The sincere emotion you are feeling
2. A brief description of the behavior you find unacceptable
3. The concrete negative effect or result of that behavior

A specific format for an I-message is as follows:

1. I feel ____________________________ (sincere emotion)
2. When ____________________________ (unacceptable behavior)
3. Because __________________________ (concrete, negative result)

For example: A new patient is present in the office today for her initial examination. You have done your first overview and have performed spot probing to determine the status of her gum tissue. Upon probing, you find evidence of pocketing and possible bone loss, so you prescribe a full mouth series of radiographs for her. She becomes belligerent and says she wants you to fix her up but she does not want "all those X-rays."

First of all, you must ask yourself the critical question, "Is what this person doing having a concrete, negative effect on me, on my performance, or on the practice?" I think you will agree that it is. Therefore, you not only have the right, but also the responsibility, to confront the issue.

For example: An appropriate I-message might be:

DOCTOR: Ms. Patient, I understand your concern. However, I feel very frustrated and helpless when I cannot obtain the necessary radiographs, because without this diagnostic information, I am not able to analyze your needs appropriately and will then be unable to recommend the therapy or treatment that will be best for you.

With this type of careful communication, your particular problem is expressed in a nonthreatening, non-judgmental manner. More than likely, the patient will not become angry, but will see your side of the situation and will realize that you are looking out for her best interest. In essence, you have stated your I-message in terms that will meet both your needs and her needs.

The opposite of an I-message is a you-message—or a put-down message—in which blame, judgment, and intimidation are the net result.

For example:

DOCTOR: *You* will not be able to receive good care, because *you* are refusing the X-rays. *You* make it impossible for us to do what we need to do.

Most of us are great at sending you-messages. You-messages are usually ineffective in getting a person to change behavior. A person will oftentimes become defensive or antagonistic. Because they have been "slam dunked," their interest in helping you is diminished.

Practice and use the I-message. This will be one of your strongest tools in handling difficult situations, whether with team members or with patients. Determine how you can confront a problem without harming the integrity of the other person. Excellent speaking and listening skills give you the foundation necessary to define problems and to be able to handle difficult people and difficult situations.

CHAPTER 7

HANDLING DIFFICULT PEOPLE AND DIFFICULT SITUATIONS

"Nobody can make you feel inferior without your consent."
ELEANOR ROOSEVELT

Have you ever had—or do you presently have—any difficult patients in your practice? Is there someone who is difficult to understand or with whom you have difficulty relating? Do some of your patients make mountains out of molehills—every visit? Are some of your patients troublemakers? Stubborn? Hostile? Do some of your patients have special needs that make treatment difficult?

If you answered yes to any of the preceding questions, then your practice probably maintains a normal flow of patients, patients that range from one extreme to the other.

Webster defines *difficult* as "hard to understand or reach, painful, laborious, troublesome, puzzling, exacting, and stubborn."

We all know difficult patients and must deal with them. Learning to deal with them effectively can turn potentially negative situations into positive situations. Excellent and effective communication in these situations can not only prevent a patient from leaving your practice and bad mouthing you across town, but it can also turn a somewhat loyal patient into a "forever" patient who becomes an ambassador for your practice.

Business experts tell us that identifying a difficult person or situation and taking action to give special care to that person or to correct the situation can be beneficial for further developing your company—in your case, your dental practice.

Let's look at some of the situations that may be causing difficulty for you, and then let's study the skills of handling those particularly difficult situations or patients. Having the armamentarium to deal with these situations will have a positive and long-lasting effect on your production, your enjoyment of dentistry, and your stress control.

DIFFICULT PATIENTS

Remember back to the chapter on understanding personality differences (Chapter 4) and know that when a patient is difficult—or your relationship with a patient becomes difficult—this doesn't indicate that either party is right or wrong; it may just indicate that differences exist. However, these differences can cause conflict and can be tough to handle.

DIFFERENCES THAT MAY LEAD TO CONFLICT OR REJECTION

In Chapter 4 I stated that two specific issues may lead to differences of opinions or to difficulties with patients. Being aware of these issues before the fact can help you offset possible difficulties and develop a greater appreciation for the uniqueness of each person.

Remember that differences that can lead to conflict or to rejection usually stem from (1) misunderstanding of or lack of appreciation for variances in personality, or (2) an incompatibility of goals.

1. Misunderstanding of or a lack of appreciation for personality differences: Many problems can be overcome if a study of and an acceptance of the variances among people is pursued. Just because a person is different from you or doesn't feel the same way as you, doesn't mean that either of you are right or wrong. It simply means that you have divergent opinions or that you are unclear about each other's goals. This lack of understanding can lead to difficulties or to fear.

Difficult behavior (including defensiveness) is often a result of fear. Being a receptive listener, letting people get their emotions out on the table, encouraging them to say what they want, understanding them and not condemning them for their ideas and opinions can relieve tension and reduce or eliminate fear.

Being a good listener and a good communicator and being willing to hear a person out lets you show respect to the person. Letting them express their own opinion without judgment will often lead to a quieter, more receptive patient, one who may become a faithful friend of the practice.

Let's review those personality styles:

- Choleric, the driving personality that wants results and control
- Sanguine, the enthusiastic personality that wants attention and positive strokes
- Melancholy, the steady personality that wants structure and organization
- Phlegmatic, the congenial personality that wants compatibility and harmony

Identifying these temperaments lets you know that each of your patients is unique and that treating each of them in exactly the same manner or presenting your recommendations in the same way is inappropriate and ineffective.

2. Incompatibility of goals: Incompatibility of goals is the source of much conflict in a dental practice. The goals *you* might be desiring may differ from the goals of the patient. On the other hand, you and a patient might want the same goal of excellent oral health, but your individual plans of action for accomplishing that goal might be incompatible. Even when conflict begins with a rational disagreement about the goal or about the action plan, emotions might erupt and get in the way of an agreeable solution.

Have you ever had a patient in the office, and no matter what transpired, it was always wrong? Did you feel that the harder you tried, the deeper you dug yourself into a ditch?

This failure to communicate is oftentimes the explanation for conflict or for unfulfilled relationships. Effective communication may not resolve all conflicts, it may not bring two people with totally different goals to a place of commonality, but using these skills can:

1. Enhance your ability to influence a person's acceptance of your recommendations
2. Smooth the actual treatment experience
3. Persuade a person to carry out necessary follow-up therapies.

What skills need to be mastered in order to communicate more effectively?

You need to be able:

- To listen so that you accurately hear and understand a person's message (Chapter 5)
- To speak so that you get your message across without closing the door to communication (Chapter 6)
- To defuse anger (Chapters 5 and 6)
- To confront constructively

The secret to getting along well with other people, even difficult people, is to determine their needs and be willing to fill them. Ask questions and listen. Ask open-ended questions, questions that cannot be answered with a yes or a no. Then listen to pick up information that might give you a clue as to (1) what motivates that person, (2) what they want, and (3) what the possible barriers to acceptance might be.

Listen to encourage a person to continue, to go on, to get it all out. Know that information is power. Ask a question and then respond with a nonstimulating response. Use excellent body language. Let them continue without interruption. Don't try to fill all silent moments. Know that silence and attentiveness can be strong communication skills.

ACTIVE LISTENING AS A CALMING AGENT

You have already learned the single most effective skill you can use to calm an irate person, to defuse anger, to handle a difficult person, or on the other hand, to enhance a good relationship: active listening. Listening is a skill that can be developed through conscious effort, and it can be improved through practice. Good listening requires effort and discipline.

Active listening is reflecting back to the person what you think you have heard them saying to make sure that you have heard them correctly. A simple repetition or paraphrasing is not sufficient. The listener should demonstrate, in his/her own words, an adequate understanding of the content, intent, and emotion of the speaker's remarks.

This type of listening gives you the opportunity to clarify, to make sure that you understand what the patient is trying to say. This type of listening has a definite calming effect by providing feedback and by maintaining a supportive atmosphere.

DEALING WITH CONFLICT CONSTRUCTIVELY

We sometimes think that it is not okay to disagree with another person, certainly not with a patient. We think that they will not like us if we express our own thoughts or feelings. And so we keep our thoughts to ourselves. Sometimes these differences of opinion, if left inside to dwell, grow from a simple disagreement to a major alienation—and even a loss of clients.

My mentor, Karen Moawad, has taught me the skills of conflict resolution. The following is my dental interpretation of those conflict resolution skills.

POSITIVE DISAGREEMENT

You can disagree with a patient in a positive manner, a manner that can prevent difficulties from developing. In disagreeing positively, you should follow these steps:

1. Actively listen to make sure that you are hearing accurately.
2. Let the person know that you understand or appreciate his/her viewpoint.
3. Express your own opinion.

For example:

PATIENT: I understand your financial policy, but I just want to pay this out.

BUSINESS MANAGER: You would rather take several months to take care of your financial responsibility, is that right?

PATIENT: Yes. I can only pay $50 per month. I can't pay all of this next week!

BUSINESS MANAGER: I can appreciate that, Ms. Patient, but we do not carry long-term accounts on our own books. If you do want long-term, convenient financing, we do offer several alternative payment options. Let's discuss these.

CONSTRUCTIVE CONFRONTATION

Do you avoid confrontation at all cost? Are you nervous about confronting another person out of fear that you will hurt their feelings, or make them angry, or cause greater conflict?

Difficult situations—difficult people—often need confrontation. Positive disagreement does not always take you far enough in the confrontation modality. You sometimes need to go further. Facing an issue head-on is the best and most constructive way in which to deal with a conflict or with problem behavior. Tucking the conflict away may create anxiety, anger, and negativity on the part of both parties. If you decide that the relationship is worth developing (or in some cases, saving), then the risk of confrontation may prove to be a gift you give to yourself, to the other party, and to the relationship. Lashing out or putting another person down is totally nonproductive. Differences handled in this manner are often unresolvable.

So how do you confront constructively? How do you let people know that you do not approve of their behavior without hurting them? How do you preserve the sanctity of the relationship and let them know that you aren't pleased about something they are doing?

The characteristics of constructive confrontation are as follows:

1. Actively listen to make sure that the feelings and the issues are clearly understood.
2. Let the person know that you appreciate their viewpoint or their situation.
3. Confront them with specific changes that must occur. Stick to concrete, tangible issues such as time, money, responsibility.

For example: You are the appointment coordinator. You have reserved two long appointments for extensive crown and bridge treatment for Ms. Patient. She rescheduled the first appointment, and she simply did not show up for the second one, even though you had confirmed the appointment the day before. Now she is calling to reschedule.

> APPOINTMENT COORDINATOR: Ms. Patient, it seems that keeping your appointments has been a problem for you. I understand that you have a very busy schedule. However, Ms. Patient, unless we can have a commitment from you that you *will* be at the next scheduled appointment, I am afraid I will not be able to reserve any further time for you. We care very much about you and about providing your treatment, but we need to ask that you respect our time so that we can provide all of our patients with our best attention. I'm sure you can understand our concern, can't you?

Notice the three parts of the confrontation. The appointment coordinator (1) summarized the person's behavior and fed that back to her, (2) let her know that she understood the challenges of her busy schedule, and (3) confronted her—very diplomatically—about the needed change in behavior.

The goal of the appointment coordinator was not to hurt the patient or embarrass her or to turn her away from the practice. Her goal was to address the behavior that was a problem, the behavior that was causing problems for her and for the whole team. She had the right and the responsibility to address this issue. Addressing it in such a way that the patient understood how her behavior was adversely affecting many other people would—it is hoped—encourage her to change that behavior. Letting her continue to schedule appointments and conveniently break them was not doing her or the practice any good.

You may be saying, "We don't need her anyway. If she doesn't respect us enough to keep her appointments, then we just need to let her go and fill her place with a patient who is more respectful." Perhaps. However, I would encourage you to try to deal with problems constructively and to try to solve those problems. Know that successful people are not people who don't have problems. They are people who have learned how to successfully solve their problems.

EITHER/OR CONFRONTATION

When extreme measures have been attempted to restore a situation or to encourage a change in behavior, but the necessary alteration has nor occurred, either/or confrontation is appropriate. Either/or confrontation gives a final solution offer, an ultimatum. It is a last opportunity to make required changes. This is an extreme measure and should be used *only* in situations where an either/or situation exists.

The characteristics of this type of either/or confrontation are:

1. Deliver an I-message.
2. Stick to the facts that are observable. Don't assume.
3. Stay in the future tense: what will happen in the future if compliance to the required change does not occur? What are the consequences of noncompliance?
4. Don't bring up issues from the past.
5. If the person tries to justify the behavior, ignore the efforts. Do not accept "yeah-buts!"

Remember, you have probably confronted this person and this issue in several *gentler* ways. Noncompliance with your previous efforts to resolve the conflict has led to this moment. Ultimately, the person is choosing this firm and powerful method of confrontation. You at this point have little or no other choice.

For example: Mr. Patient comes to the office for treatment that will take several appointments. He continually speaks in a suggestive, inappropriate manner to the ladies on the team. Today he makes a pass at your clinical assistant.

> DOCTOR: Mr. Patient, I feel distressed when one of my employees is taken advantage of by one of our patients, because that is unacceptable, inappropriate behavior. Today you grabbed Jan and upset her. I respect my team members and expect the same respect from our patient/clients. Either you refrain from handling Jan—or any of my other employees—or I will be forced to excuse you from my practice.

This is a strong measure, a strong method of confrontation. Perhaps you will never need this type of confrontation knowledge. But, if you do, this is a proven method of handling difficult situations and difficult people in an acceptable manner.

CONFLICT APPRAISAL

The process of resolving a conflict with a difficult patient is a complex process with consequences that endure beyond the event itself. Conflict is inevitable. Thus an understanding of and a competence in the skills of effective communication and conflict resolution are critical.

Ask yourselves these questions:

1. Understanding the level of difficulty, is it worth the effort it will take to deal with the person?
2. Are we willing and able to find out what this person's particular needs might be?
3. Can we meet those needs?
4. Are we beating our heads against the wall?
5. If the conflict is not resolved, if we do not come to equitable terms for both parties, do we want to maintain the relationship?

It will take an incredible amount of effort to put the needs of the difficult patient in front of your own. They may not realize your efforts; they may not appreciate your efforts at all. Ask yourself the previous questions. Answer them honestly. Do all you can do to handle your difficult patients.

Spend time as a team studying, discussing, and practicing confrontation skills. Then commit to putting the skills into effect for the betterment of your practice. Handling difficult patients/difficult situations can be done. And it can be done in a manner that will control the stress produced by these difficult situations.

Do We Create Our Own Difficult Patients?

I once spoke with a 37-year-old doctor who was in his 11th year of practice. He was calling to discuss my consulting services. His practice was not doing well, and he was extremely frustrated.

During that initial interview, some obvious barriers to his practice success emerged:

1. He didn't like doing dentistry!
2. He was in financial woe.
3. He said he had no systems in place for management. He had no systems at all!
4. He resented his patients. He felt that he was performing a well-rehearsed play every day, because he really didn't like what he was doing and didn't even want to see those patients. He said that dentistry for him was only a means to an end: money (of which he had none).
5. He was attracting very few new patients to the practice and was concerned that the end was near.
6. He was a graduate of the most respected post-graduate clinical institutes but felt that he had been misled by all of the "gurus" who believed that commitment to quality could and would lead

to personal, professional, and financial fulfillment. So he was angry.

7. He felt that everyone was out to get him.
8. He said his greatest frustration came when people wanted the high-quality, long-lasting, proven service, but were angry at him for the fee.

REFLECTION: THE MIRROR EFFECT

Don't our patients reflect us! This conversation with this frustrated doctor validated an underlying philosophy of my management: that patients are a mirror of you and of your practice. If you don't like how your patients respond to you or to your recommendations, put a mirror in front of all aspects of your practice, from telephone to greeting to presentation to financial arrangements to treatment to follow-up. What are you putting out there? Whatever you are putting out there will come right back to you.

MOTIVATION

Harvard University researchers wanted to know what drives a person to go beyond the average, to become fantastic in their work environment. They found that six factors are primary motivators in the workplace today.

The following factors in order of strength characterize the highly motivated individual:

1. Achievement
2. Recognition
3. The work itself
4. Responsibility
5. Advancement
6. Personal and professional fulfillment

Let's see how the presence or the absence of these factors may have negatively affected this young doctor.

Achievement: He wanted to achieve clinical expertise. He had attended the best clinical courses and numerous management courses. However, the achievement of this goal—clinical excellence—had been stifled by his inability to attract people to his practice and by his inability to gain acceptance of treatment recommendations.

He wanted to achieve financial security, but this goal had not been accomplished because a healthy new patient flow and case acceptance had not been achieved. One ties directly into the other.

The cycle had become a vicious one. The fewer the number of new patients, the fewer cases presented. The fewer cases presented, the fewer cases accepted. The fewer the cases accepted, the lower the revenues for the practice. The lower the revenues for the practice, the more discouraged he became. The more discouraged he became, the fewer the new patients, etc., etc., etc. This had become a vicious cycle for him. What you put out there in all things—including attitude—comes back to you.

VICIOUS CYCLE—THE MIRROR EFFECT

Recognition: Recognition is the number two motivator of people in the work environment. This doctor said:

> *"I'm over wanting to be recognized as a quality dentist—or as the practice that offers 'only the best,' etc. I just want to make a living. This dentistry stuff is just a job. It's just a means to an end."*

Tied closely to recognition by colleagues or clients or family is a strong sense of well-being, a strong sense of self-worth. If a person doesn't feel good about him/herself, if a person doesn't believe in personal worth, if a person doesn't love him/herself, then it is impossible to love another fully, sincerely, unconditionally.

Here, perhaps, was the greatest barrier, the greatest problem for our young doctor. This man couldn't love his patients, or his staff, or his practice, or dentistry. His sense of self-worth was so low that he had lost the level of caring for himself. He needed to feel good about himself before he could reach out to others. If he couldn't sincerely reach out to others, who would be attracted to the practice? If he wasn't confident in himself, how could his patients be confident in him? Without this strong level of confidence and trust, case acceptance had become an unreachable goal.

VICIOUS CYCLE—THE MIRROR EFFECT

The work itself: Earl Nightengale says, "We become what we think about." The subconscious mind does not know the difference between reality and nonreality. Therefore, whatever is fed into the mind stays there *forever.* All thought processes are filtered through the subconscious mind.

Therefore, when this doctor constantly said to himself and to others, "I don't even like dentistry," his mind acted on those thoughts. The mind goes to work to put thought into action.

Not liking the dentistry became a self-fulfilling prophecy. The more he thought, "I don't like dentistry," the more he didn't. The more he didn't like dentistry, the fewer patients he was attracting, the fewer new patients he was attracting, the fewer cases accepted, the fewer the number of cases accepted, the lower the practice revenues and so on.

VICIOUS CYCLE — THE MIRROR EFFECT

Responsibility: Human beings thrive when entrusted with responsibility. This entrustment is a motivator. Enthusiasm for a project is generated when responsibility is given and trust is imparted.

When another person gives us the responsibility to provide a service or to carry out a specific task, we are motivated. The human being responds more effectively to positive reinforcement than to negative reinforcement. Entrustment and responsibility is perceived as just that: positive reinforcement.

This doctor was experiencing a void in his inner self, because he was not being given the responsibility of caring for many patients. Most doctors tell me that the driving force for entering dentistry in the first place (and the sustaining drive for continuing in the profession) is having the opportunity to help people achieve greater health and optimum beauty. If that responsibility is not an ongoing part of the work itself, then motivation exits—and love of the work itself leaves.

VICIOUS CYCLE — THE MIRROR EFFECT

Advancement: In the corporate world, the opportunity for advancement is obvious. One knows what opportunities lie ahead and where one can go in the advancement scenario.

But in dentistry advancement doesn't have the same black-and-white definition. Advancement is less obvious. However, advancement for the dentist might mean:

1. Financial advancement (more net profit)

2. Greater clinical proficiency

3. Leadership development

4. Team building

5. Addition of associates/partnerships

Our young doctor friend was not accomplishing any of the above. Therefore, advancement was *not* a motivator. In fact, the lack of advancement in each of these areas was becoming a *demotivator.*

VICIOUS CYCLE — THE MIRROR EFFECT

Personal and professional fulfillment: Money is a motivator. But it will only go so far with the human being. We all know of a person who has everything except happiness.

Our beloved and respected mentor L. D. Pankey reemphasized Aristotle's Cross Of Life to those of us in dentistry and encouraged us to strive for a balance in our lives—a balance between work/play/love/worship, shown in Fig. 7–1.

You can and will accomplish personal and professional fulfillment in the practice of dentistry when you realize that the balance gives energy and motivation and happiness. As simple as it may sound, that's what each person is really striving for in dentistry and in life: happiness.

Fig. 7–2, another cross from Dr. Pankey, indicates a further area for needed balance: knowing yourself, your patients, your work, and your knowledge. Personal and professional development depend upon this balance.

Combine these four intricate and essential factors starting with

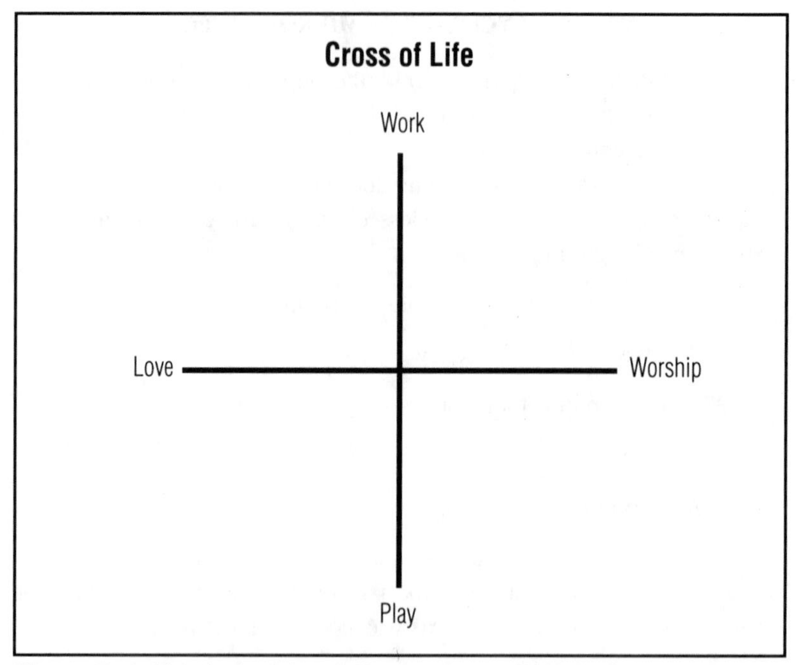

Figure 7–1 This is the Aristotle/Pankey Cross of Life indicating balance and happiness.

number one: Know yourself. Knowing yourself—the good and the bad—is essential for growth.

PUT UP THE MIRROR

Put a mirror up to your practice and to yourself. Take an honest look. Find the strengths; maximize these. These are your gifts. Use them well. Find the weaknesses; work on overcoming these weaknesses. The mirror effect is strong. The strengths *can* be turned in your favor. Concentrate on the positives. Turn negatives into positives. Feed your mind the good stuff—and that's what will come your way. Remember: "You become what you think about."

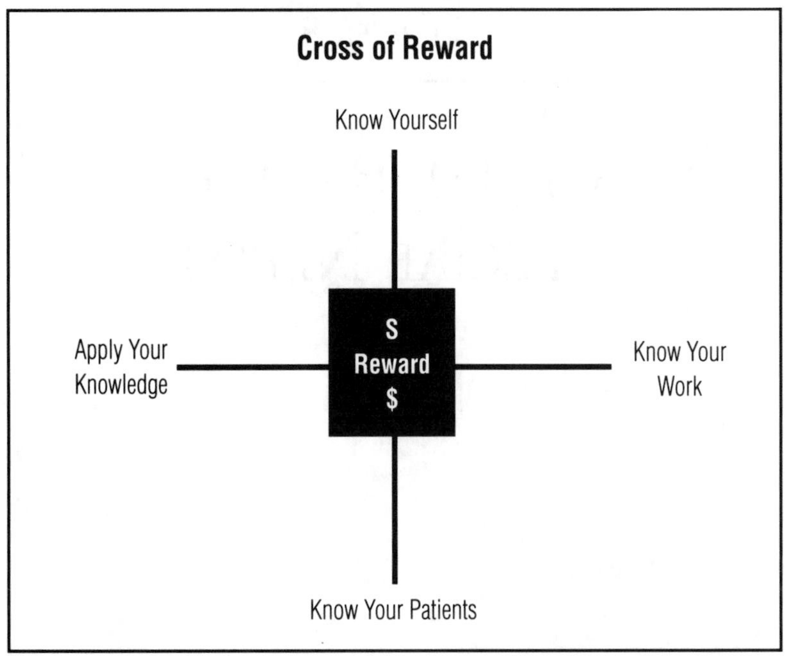

Figure 7–2 Dr. Pankey suggested this Cross of Reward as a source of professional fulfillment (S= spiritual reward, $= financial reward).

Defining your problem or weak areas is the first step toward solving the problem or overcoming the weakness.

All of us must learn to stop looking *outside* of ourselves for something to blame for our difficulties. We must begin looking within.

I have spent many hours in deep thought about this young doctor. His challenges are great. However, the acceptance of challenge may be the greatest motivator of all.

CHAPTER 8

SPECIAL PEOPLE/ SPECIAL NEEDS

THE PARENT/CHILD

Providing dental care for children can be an extremely enjoyable and rewarding experience. As dental professionals and as dental educators, you must be aware of and use methods for building strong relationships with the children and their parents before treatment can be successful. The ability to deliver quality treatment in a nonstressful manner depends upon the cooperative behavior of the child.

The parent or child—or both—become difficult patients when *fear* becomes a dominating factor in the dental experience.

Children can develop fear in one of three ways:

1. Learning it as a direct imitation of a family member (brother, sister, or parent) or from a friend
2. Associating the dental experience with an existing fear
3. Actually having a painful or frightening experience

If an appointment with a child becomes uncontrolled because of fear—if the child tries his/her very best both physically and vocally to protest the impending treatment—then both the child and the dental team suffer. C. M. McElroy says, "Although the operative dentistry may be perfect, the appointment is a failure if the child departs in tears."

Experts in the area of dental behavior agree that dental attitudes usually develop in childhood. The development of a negative attitude caused by a bad experience in childhood can lead to dental fear and dental avoidance in adulthood. Therefore, the best way to handle the difficult parent/child situation is to prevent the situation in the first place whenever and wherever possible.

PREVENTING FEAR

As dental professionals you are committed to prevention—clinically. In addition, being committed to prevention behaviorally serves both the patient and the provider. Preventing or constructively dealing with fear will lead to positive relationships between the parent and child and the dental team.

What can be done to set the tone for a good parent/child situation?

Environment

A child is acutely aware of his/her surroundings—every aspect. The senses are tuned in to everything that is happening as a result of the child's innate inquisitiveness. The child's interpretation of the dental experience is affected by all of the senses: hearing, sight, smell, touch, taste.

Hearing:

- Noises made by the dental equipment: suction, handpiece, amalgamator
- The sound of the doctor's or the assistant's voice: quiet and smooth is relaxing, while hurried and rough can cause tension
- Words: use of positive words can set a positive tone, while use of negative words can stimulate fear or apprehension.

For example:
"make you comfortable" vs. "get you out of pain"
"special pictures of your teeth" vs. "X-rays"

Sight:

- Have a specially designed space for the children—a space in which they feel at home, in control, and comfortable.
- Games, books, children's magazines, video games, small tables—all of these give the appearance and the feeling that the child is welcome in your practice. They don't feel so overwhelmed by *big* things!
- Smiling faces from all members of the team—including the doctor—will calm a child.
- Instruments and needles should be kept out of the sight of the child.

Smell:

- Try to soften any dental smells by using air freshener or fresh flowers (kept out of the reach of the children, of course).
- Scented candles can be burning in the business office so that the soft fragrance wisps throughout the office.
- Use the scented latex gloves that are now available.
- Careful attention must be paid to the grooming of the dentist and the assistants. Be sure to be pleasantly scented. Include the use of mouth wash on a regular basis throughout your day.

Touch:

- Touch the child with firm, gentle touches of reassurance.
- Let the child hold the saliva ejector in his/her mouth. The child will love being your "helper" and will be more cooperative.
- Allow the child to touch the equipment and instruments: the mirror, the handpiece, the hoses. Dr. Bill Bozalis, a graduate of the pediatric department of Northwestern University Dental

School says that he lets the child hold the mouth mirror. Dr. Bozalis says, "If children feel that they are in some control, they are more cooperative."

- Shake hands with the child. This will give the child an important physical contact with you and will make him/her feel special. Dr. Bozalis shakes hands with the child while he is gloved. Then he lets the child feel the gloves. Of course, no instruments are in his hands at the time.

Dr. Bozalis explains,

> *"We let the kids feel the gloves, the mask, etc., so that they are not intimidated or alarmed at the infection control tools. They realize that they are a normal part of the visit. Once they have had a chance to feel the tools, they seem less concerned about them."*

Taste:

- Make sure that the topical anesthesia has a flavor that the child will like.
- Rinse the child's mouth quickly to eliminate any unpleasant tastes.
- Fluoride and toothpaste flavors must also be pleasantly flavored. You may wish to offer a variety of flavors to give the child that important sense of choice and control.

Preparation

In addition to providing the criteria for setting the tone for a positive dental experience for the child, carefully preplan the visit. It will be to everyone's advantage. Dr. Eugene Litteken, a graduate of Baylor University School of Pediatric Dentistry, believes that educating the parent and the child prior to and throughout treatment is "the single most important factor for effective treatment to take place."

Dr. Litteken believes that one of the most difficult situations with which he has to deal in his pediatric practice is when a parent or a child comes to the office with misinformation or with an incorrect interpretation of information.

> *"Parents or children may tell you one thing, but what's really happening is something totally different. What they are saying and what they are meaning may also be two different things. By asking open-ended questions that establish good lines of communication, we are able to see where the parent and child are coming from. In this manner, both feel a sense of comfort and a sense of control. They know that we are listening to them and that we want to work with them to solve their problem or to meet their needs. Then they will be more likely to cooperate with us when we recommend treatment. That initial contact is critical."*

The preparation of the parent and the child for their initial visit is essential for a cooperative new-patient experience and for a long-term relationship. Much effort has been placed on figuring out what type of initial experience will be most likely to reduce anxiety.

How the parent prepares their child for the first visit to the dentist will often set the tone for that visit. The dental team needs to take the time and put forth the effort to educate the parent in advance of the appointment. This will reduce the risk of the parent negatively preparing the child. Children who are well prepared for their visit tend to be more cooperative than those children who are inappropriately prepared.

Parents who themselves are not dentally literate or who have had a negative experience often provide a poor preparation for their children. Parents who do not really know what a dentist does or how he/she does it will often mislead a child, sometimes in a harmful way.

For example: A parent who thinks a dentist is someone who pulls teeth will not likely prepare a child for a happy visit. When children go into an appointment thinking something is wrong, they are less cooperative. However, when they go into the appointment thinking that the dentist is someone who "takes good care of you and of your teeth," they are more relaxed, and thus more cooperative.

The preappointment letter

The preappointment letter has been recommended for several decades as an effective manner in which the dentist can help the parent to properly prepare the child for their initial visit. Research has supported the concept that such an educational piece when sent prior to the initial visit will do the following:

- Help the parents to be better prepared,
- Encourage the child's cooperative behavior, and
- Reduce broken appointments.

This educational letter does not need to be long or complicated. It should serve as a welcome to the parent and child and should provide a brief description of the initial visit. It should give the parent some suggestions on how they can prepare their child for this visit, including some verbal skills for describing the dentist and the visit (see Fig. 8–1).

In addition to the letter sent to the parents before the visit, we recommend sending the child a card or note addressed specifically to him/her (see Fig. 8–2). Children, for the most part, don't receive much mail, but are usually thrilled when they do. Receiving a note from the dentist before the visit may serve to relieve anxiety and to instill some enthusiasm for the meeting.

The initial visit

The initial visit should consist of the steps listed in the preappointment letter. Children, like adults, fear the unknown. On the other hand, they gain comfort and security from knowing what to expect. Therefore, the classic tell, show, do method needs to be used consistently with children—and adults!

1. Use simple, straightforward language when addressing the child. Explain what you are going to do before each procedure.
2. Demonstrate the procedure on yourself or on an inanimate object.
3. When you are confident that the child understands what is going to be done, then proceed.

Dear // (parent's name)//,

Welcome to our dental practice. All of us on the team look forward to meeting you and your child (children). We are committed to taking good care of both of you.

We wanted to let you know what will be happening on that first visit. This information will help both of you to feel more at home and more comfortable with us.

We want to help your child be an excellent dental patient who will be able to accept routine dental care. Preparing the child at home, prior to the initial visit, is very important. The following are some suggestions to guide you as you prepare your child:

(1) Your child's dental visits are going to be a normal part of growing up. Please do not give your child the idea that there is anything to fear. There isn't.

(2) Don't make a big deal out of the visit. It is best to tell your child the day of the appointment.

(3) If your child should ask questions, explain that the dentist will look at his teeth to make sure they are healthy.

(4) Please do not threaten the child with a visit to the dentist for anything—misbehavior, not brushing, eating the wrong thing, etc.

We are enclosing a medical history and a patient information sheet for you to complete and mail back to our office in advance of your visit. Filling this out at home will let you do so at your own pace and will allow us to see your child more quickly upon arrival. We have included a self-addressed, stamped envelope for your convenience.

During your child's first visit, the following things will happen:

(1) We will review the child's health history with you.

(2) We will spend time with the child to get acquainted and to assess his/her emotional development.

Figure 8–1 Set the tone for a positive experience for your child patients by sending this informative letter.

(3) We will try to build a comfortable relationship with your child by asking questions so that he/she is involved with the appointment. We want to know your child better, and we want him/her to know us better.

(4) We will discuss any areas of concern with you.

(5) We take a look at the child while he is sitting up to assess dental conditions.

(6) We take special pictures of the child's teeth.

(7) We keep this first visit very low-key and ALWAYS tell the child what we are going to do before we do it. We TELL the child what is going to be done. Then we SHOW him what is going to be done. Then we DO it.

(8) We will always tell your child when even the slightest thing goes well. We want to positively reinforce your child to establish his/her good behavior.

(9) Then we schedule an appointment for a cleaning.

(10) If further treatment is necessary, we will discuss this fully with you prior to the making of an appointment. There will be no treatment provided until we have a strong relationship built with the child. This will be important for your child's continued comfort as a dental patient.

Our practice is committed to prevention and to total health. By starting your child early in life, we can prevent decay and dental disease with early detection, oral hygiene care, and diet counseling.

Your child, with your help and cooperation, can become and remain a good dental patient with a healthy mouth and a happy smile.

We look forward to meeting you.

Sincerely,

Dr // // and Team

Figure 8–1 *(Continued)*

Dear ________,

Our dental team is so excited that you will be coming to have your teeth cleaned and checked on ________ at _____.

It is very important to begin early in your life to learn how to take great care of your teeth. We will work together to help you keep happy, healthy teeth for a lifetime.

See you soon,

Figure 8–2 Make your child patients feel special by sending them a personal letter (courtesy of Dru Halverson, RDH, Jameson Management Group).

As important as the tell, show, do method has proven to be throughout the years, nothing seems to be more significant in helping to assure a positive dental visit for a child than friendly interaction with the dental personnel outside of the operatory environment. Research has shown that this type of positive interaction outweighs any other type of preparatory efforts or explanations.

Therefore, everyone on the team must realize how significant their interaction with the child may be! Time spent with one of the dental auxiliary in a neutral area of the office seems to be one of the strongest criteria for a calm, relaxed, confident appointment for the child.

Learning Models

Modeling: Permitting a child to watch other children or to watch Mom or Dad undergo dental treatment has been recommended as a method of preparing a child for their initial dental experience. This method of learning states that children learn by watching. Show children what they can expect to happen and how they are expected to behave. Children do not know what is expected of them in the dental environment, and even though you can explain the required behavior, nothing is more effective than for them to actually see what is expected of them.

Modeling is one of the most accepted and proven methods of eliciting calm, cooperative behavior in the child—especially the dentally inexperienced child. If a cooperative sibling is available for demonstration, the neophyte child should be given the opportunity to observe this sibling undergoing treatment. If a sibling or a parent is not available, another cooperative child can be observed. (Of course, you will want to make sure that the chosen child or model is cooperative.)

Distraction: The length of time the child is asked to sit in the chair for treatment depends on the child's ability to remain immobile. It is normal for children to have short attention spans and short spans of time in which they can sit still. Vary the length of time for the appointments according to the child's own physical and emotional abilities.

If you are performing longer appointments, distraction may be to your advantage. Videotapes and mounted TV monitors can provide such distraction. Headsets with music or stories can also prove to be a valid distraction. Time will fly by for the children, and they will be more cooperative if they are thinking of something besides what you are doing to them.

Reward and punishment: Consequences following a specific behavior play an important role in learning. It is important to understand the effects of both positive and negative reinforcement.

It is important never to punish a child in the dental environment.

Physical punishment has no place in the dental arena. Mild punishment is usually ineffective in the first place, and severe punishment can lead to a lifetime of dental phobia.

Albert Bandura states in his research on this subject,

> *"The resulting avoidant responses may be more socially undesirable than the behavior the punishment was originally intended to reduce, and once established, these behaviors may be considerably more difficult to eliminate."*

Punishment must be administered early in a sequence of behaviors to be effective. The administration must be consistent. If children are not punished until they have full-blown tantrums, then the effectiveness of the punishment is greatly reduced. To be effective, intense punishment must be administered for each instance of uncooperative behavior. This is obviously not acceptable.

Furthermore, research on the use of punishment has shown that more often than not, punishment has the opposite effect than might be desired. It usually increases uncooperative behavior, the tantrum, and aggressive behavior. Efforts to control a child's behavior with verbal or physical force result in children who are resistant and uncooperative.

On the other hand, when the dentist gives simple instructions to the child as to what is expected, and then gives the child feedback on their performance, cooperation usually follows. An instruction-only method of learning solidifies cooperative behavior far better than a punishment method.

Communication

Dr. Litteken says,

> *"It is absolutely essential that we establish open lines of communication with the child if we are to be successful with future care."*

In order to establish those open lines of communication Dr. Litteken and his team use the tell, show, and do method as previously outlined. Says Dr. Litteken,

> *"We educate our parents and our children about our standard of care. If they tell us it has been done differently in another office, we tell them that we understand this, but in our office we do it this way. Then we explain why—in terms of how the difference will be beneficial to the child.*
>
> *"We find out what the problem is and then design a plan of action for the child. We try to keep the parent involved with the design of the plan of treatment. We do this by asking open-ended questions of the parent. This makes them feel comfortable and gives the parent a sense of control. When the parent is given the option of making choices, they sense control—even though the doctor is actually in control."*

Assuming this type of respectful communication modality with the parents is critical. It is also critical that the entire team address the child with the same type of respect. Speaking in a normal tone of voice with normal inflections is more effective than talking down to the child. Sometimes people change the way they speak when they are with a child. Artificial speech, a high-pitched tone of voice, a sing-song rhythm of speech, or "ooey gooey" language are inappropriate. Likewise, speaking slowly and specifically as if this would allow the child to understand more clearly is also an inappropriate method of communication. Children will respond most effectively when you use the same tone of voice with which you address your adult patients.

Dr. Norm Olson, the dean of Northwestern University School of Dentistry, encourages the dentist and the entire dental team to "Treat the children like adults and the adults like children."

Food for thought, wouldn't you agree?

VOCABULARY

In terms of vocabulary, a child—depending on the age—will have certain limits of understanding. Be sure to address your education and your explanations of treatment in terminology that is concrete rather than abstract, present tense rather than future tense, relevant to the child rather than relevant to another. If you make a promise, keep it. If you tell a child something, mean it. If you ask something of a child, follow up on your request.

Be honest and open with your children patients. Let them know what you are going to do before you do it. Let them know what you expect of them, inform them of the expected mode of behavior. Give the child feedback—positive feedback—throughout the procedure.

SPECIAL BEHAVIORAL PROBLEMS WITH CHILDREN

The shy child: Do not force yourself on the shy child. Do not try to tease the child into being more responsive. This will usually produce the opposite of what you want. This effort to bring the child out will usually drive him/her deeper into Mommy's arms. Focus your attention and your comments on the parent. Then from time to time, look at the child and give him/her a wink, but make no verbal contact with the child at this point. At first the child will withdraw, but sooner or later will be looking toward you, anticipating the wink and the attention. Before long, you will be able to draw the child out by asking questions and by getting him/her involved in the conversation.

Avoid teasing a child about a state of anxiety. Doing this will only prove to the child that—just as he/she thought—it is not okay to be afraid. Telling a person that it is not okay to be afraid will do nothing to defuse that fear. Rather, acknowledging the fear, letting the child know that you understand his/her feelings, but that you are there to help, will go much further than cajoling or teasing the child about the fear.

The out of control child: All of the techniques discussed heretofore are geared to the prevention of uncooperative behavior. Practiced with consistency, these methods of interaction are proven effective. However, there is a small segment of the child population that is, in fact, out of control and very difficult to handle.

Children who exhibit out of control behavior oftentimes begin to throw a fit before they enter the treatment area or as soon as they enter the treatment area. The fit accelerates as the dentist enters the environment and increases as the doctor begins—or tries to begin—administering treatment. These children do not whimper, they howl. They do not squirm, they kick and fight. They do not protest, they try to escape.

This type of situation can prove to be trying, even exasperating for the most even-tempered doctor in the world. Don't think that you must win the battle at any cost. The cost could mean a lifetime of dental phobia for the child and psychological trauma that will be very difficult—if not impossible—to overcome. This type of no-win situation will often leave the dentist feeling guilty and unprofessional. If the child cannot be handled properly and comfortably, stop treatment and contact a specialist for advice and/or referral.

The protesting child: This is the child who is resistant to all that you ask of him/her. They protest any procedures or any requests for cooperation.

Dr. Bozalis indicates that three difficult situations posed by the defiant or resistant child are:

1. Making too much noise—screaming, hollering, crying
2. Refusing to open their mouth
3. Not sitting still during treatment

Efforts to coax, to bribe, or to use adult logic are usually of no use.

In some situations reverse psychology is effective. To use this technique, begin by actively listening to the child's behavior. Reflect back to the child what you perceive him/her to be feeling.

For example: "You don't want to let me look at your teeth, do you?" or "You are feeling angry, and you want me to know that you are angry, don't you?" You will usually grasp the child's attention with this type of reflective listening.

At this point begin implementing the reverse psychology by encouraging the child to go ahead with the behavior.

For example: "Show me how mad you can become." "Throw a bigger fit." "Hit the chair." "Stick your tongue out at me." And so on. The child will usually go ahead with the prescribed behavior, but will quickly tire of this because the behavior is not getting the desired result. The dentist isn't getting angry at all, but is encouraging the performance. Some children will find the situation humorous and will end the situation by laughing at the situation and at themselves. Continue this scenario until the child is ready to depart from the defiant behavior and enter a more cooperative state.

When to Refer

In determining when a referral should be made, consider these three guidelines given by Dr. Litteken to his referring general practitioners:

1. Know your limitations.
2. Know your level of tolerance.
3. Be able to assess a situation objectively. Stop or make changes, if necessary.

When a general practitioner is referring a child to a pediatric specialist, a history of the child needs to be given. This history should include such information as:

1. What has been comfortable for the child?
2. What has not been comfortable?

Describe both the clinical and the behavioral history of the child. This information will help the pediatric dentist and the team to be better prepared for the child's arrival.

MANAGEMENT OF CHILDREN PATIENTS

Children can be the most difficult of all your patients. And, they can be one of your greatest sources of joy. Study this chapter and look for concepts you can apply to your own practice. Then begin to implement the ideas on a regular basis until they become a natural part of your patient management format.

Concentrate on the prevention of behavior problems just as you concentrate on the prevention of disease. The effects of a negative dental experience during childhood can be as ravaging as tooth decay! When you make a concerted effort to prevent fear in each young patient, you will be providing a lifetime gift to that person.

Develop a level of confidence with your children. Your confidence and the confidence of your team will have a powerful and reassuring effect on your young patients. Maintain self-control at all times. Give your children your sincere love and attention. They will mirror that love and attention right back to you.

DO'S and DON'TS

DO'S

1. Compliment good behavior. Give positive reinforcement.
2. Give specific instructions.
3. Ask open-ended questions to discover the child's feelings.
4. Plan a great initial visit.
5. Ignore negative behavior.
6. Give the child comforting physical contact: pats, strokes.
7. Gradually proceed with treatment based on the child's readiness.
8. Have short appointments—or appointments that fit the child's tolerance level.
9. Control yourself and your temper.
10. Involve the child when possible. Let him/her be a helper.
11. Carefully evaluate your facility, your appearance, and your systems.
12. Send a preappointment letter.
13. Refer when appropriate.

DON'TS

1. Treat children if you have no tolerance for them.
2. Bribe a child.
3. Force the child or make threats to the child.
4. Ignore the child.
5. Humiliate the child or put him/her down.
6. Use physical force.

COMMUNICATING WITH THE GERIATRIC PATIENT

Our *daily* input of media from television, telephone, computer, and other sources is more than our grandparents received in a lifetime! Therefore you must not expect your geriatric patients to meet you where you are coming from, but rather *you* must meet *them*. Here are some suggestions for assisting your geriatric patients:

1. When a nursing home patient enters the office, the person administering the business office needs to *go in to* the reception area to assure the comfort of the patient.

2. Provide easy access to your facility.

3. If this person is a new patient, the necessary paperwork needs to be sent to the nursing home before the patient's visit, so that the attending staff can complete this paperwork while the records are in front of them.

4. If for some reason the necessary paperwork has not been completed, stop what you are doing and go to the reception area to assist with the completion of the data.

5. Depending on the age and hearing ability of the patient you may need to slow your speech somewhat, repeat what you have said, and speak a bit louder.

6. Assist the patient into and out of the dental chair. Gently tell the patient you will be lowering the chair *before* you lower it!

7. Touching is vital to people! Firmly and gently touch these wonderful individuals to let them know you care, to give assurance and security, and to add a sense of warmth to their experience with you.

8. Treat these patients with extra attention in a mature, respectful, *adult* manner.

9. Always give the patient the opportunity to spend a few moments in the restroom (assisted, if necessary).

10. Along with the nursing home attendant or family member—whoever brings the patient to the office—escort the patient to the front door or to the car. The value of this extra attention, courtesy, and assistance is immeasurable.

In Summary

All patients need and deserve excellent and special attention, but these folks need a little extra. You could be a bright spot in their lonely days. You will give extra attention to them, but you will be the ultimate receiver.

Chapter 9

Communication By Telephone

"The telephone is the most important marketing tool you have in your practice. Give it the respect it deserves and the attention it needs."

Dr. John Jameson

Your telephone is your artery to the world. Conversations held on the telephone make a statement about who you are, what you do, and how you take care of your clients/patients. This is why all members of the team, from the dentist to the newest auxiliary, must know the office's telephone protocol.

Any time a member of the team picks up the telephone for either an incoming or an outgoing call, that team member must focus on the business at hand, which is:

1. Greeting the person on the phone
2. Establishing the need of the caller

3. Listening attentively

4. Responding to the determined needs

Answering the telephone should be a high priority. When the telephone rings, answer it between the second and third ring. Follow the suggestion of Bell Telephone and answer the telephone in the following manner:

"Good morning. Dr. Jameson's office. This is Cathy. How may I help you?"

By greeting the person in this manner you serve several purposes:

1. You give the person a pleasant, warm, enthusiastic greeting offsetting, perhaps, a negative attitude about calling the dentist.

2. You let the caller know that they have reached the correct number.

3. You give them your name to attach to the voice so they can begin to *bond* with the practice and the people therein.

4. You have asked them if you can help—and, after all, that is what you are there to do.

If the telephone rings and you are with a patient, courteously excuse yourself and answer the phone. If in establishing the caller's need, you discover that you will need a few minutes to respond to them, ask if you may have a name and telephone number to call them back. If you do this, be *sure* to call them back!

For example:

TEAM MEMBER: Mrs. Jones, I am going to need a few minutes to discuss this insurance issue with you. I am with a patient at the moment. May I take your telephone number and return your call?

MRS. JONES: That would be fine. My number is 555-6011.

TEAM MEMBER: That's 555-6011. (Repeat the number to assure accuracy.)

MRS. JONES: That's right.

TEAM MEMBER: Thank you, Mrs. Jones, I will return your call in (give a reasonable range) minutes. I appreciate your understanding.

If you have *almost* completed your tasks with the patient at the desk, you can ask the caller if you may place them on hold. Notice that I said ask the caller if you may place them on hold. People usually don't mind being placed on hold if they are asked rather than told.

For example: "Mr. Sanders, may I place you on hold for just a moment? Thank you."

If he says no, take a name and telephone number and call him back *immediately*.

When you do place people on hold, pick up the phone every 30–60 seconds, and let them know you haven't forgotten them.

For example: "Mr. Sanders, I'll be with you in just a few moments. Okay?"

When you do return to give your full attention to the caller, thank the person for being patient.

For example: "Mr. Sanders? Thank you for your patience. How may I help you?"

When you are (1) taking messages or (2) gathering patient information from a telephone call, correct information is critical. Each of these types of telephone communication is so vital that each deserves specific attention.

TAKING MESSAGES

When an incoming caller asks for the doctor or a member of the team, the general policy should be the following:

1. The status of the person being called should be given (is he/she busy or not?).
2. If he/she is busy, a name and phone number is taken on a message pad like the on in Fig. 9–1 that has duplicates.
3. The appropriate message is taken.
4. The information is confirmed.
5. The message is placed in a designated location for convenient retrieval.

For example:

TEAM MEMBER: I'm sorry, Mary is with a patient at this time. May I take your name and telephone number? I will ask Mary to return your call as quickly as possible.

That's Jack Smith (get last name also) at Ace Office Supply, 555-4210. Is that correct?

Thank you, Jack. I'll give Mary this message.

Be sure to repeat the message to clarify and to make *sure* that you have taken the message correctly. Then make sure she gets the message! Have a specific bin or a specific bulletin board where messages are placed.

Telephone Message Pad

Message

For: ______________________

Date: ______________________

Time: ______________ A.M.

P.M.

Caller: ______________________

Company: ______________________

Telephone: (   ) ______________

Ext: ______________________

Message: ______________________

____

____

Figure 9–1 Carefully gather necessary information from *all* telephone calls.

This scenario serves the following purposes:

1. You've graciously let the person know that the recipient of the call is unavailable, but that they can count on you to get the message to her.
2. The name and number of the caller have been gathered, and you have repeated this to make sure the information is correct.
3. You have a specific place to post messages so that the message is received. If this does not happen, both parties lose.
4. You've applied professional management to a vital area of the practice.

Personal telephone calls must be limited. These calls can be very costly to the practice in terms of time and money. Personal phone calls cause the following:

1. Lost production time for the business as well as the clinical staff
2. Interference of the schedule that can negatively affect many people
3. Long distance bills to the office (in some practices this applies)

Some calls are passed on to the doctor. However, the doctor needs to let the business administrator know which calls will be received. (i.e., specific referring doctors). Otherwise, apply careful screening and monitoring of all calls.

GATHERING PATIENT INFORMATION

The telephone information pad in Fig. 9–2 is an example of the type of information that needs to be gathered when a potential patient calls your office. When a person calls and you do not recognize the name, ask, "When was the last time you saw Dr. Jameson?" Rather than, "Are you a new patient?"

You try your best to remember people, but if you are new to the practice—or if you have a slip of memory—you may forget a person's

Patient Communication Slip

Date: _______ Name: _______________ (Pronounced) _____________

Adult: ________________ Child: __________________ Age: ________

Family members who are patients: ___________ or none: _____________

Referral source: ______________________________________

Purpose of call: ☐ Emergency ☐ Examination ☐ Other

Comments: ______________________________________

____

____

Address: ______________________________________

Home Phone: ____________________ Work Phone: _______________

Dental Insurance: ☐ Yes ☐ No

Carrier: ______________________________________

Employer: ______________________________________

Appointment Set: ______________________________________

Figure 9–2 Gain insight about new patients by collecting information at the initial telephone call.

name. But callers may be offended if they've been with the practice for a while and you ask if they are a new patient!

As soon as the person gives a name, write it down on the telephone information pad. If the name is somewhat unusual or difficult, write a phonetic spelling of the name and repeat it to make sure you are pronouncing it correctly. People like to have their names spelled and pronounced accurately. Then ask appropriate questions throughout your conversation and fill in the information. Repeat the person's name several times throughout the conversation. This gives a person a special sense of priority with you.

Having this information at the outset gives you a head start in preparing the necessary paperwork for the person's arrival to your office. It also gives you a common base upon which to build rapport with this new patient at their initial visit.

TELEPHONE COMMUNICATION: THE BOTTOM LINE TO INCREASED REVENUES

Enhanced telephone communication can increase the revenues of your practice. Let's look further into the value of telephone communication as a practice builder and as a way to increase production.

"Cathy, I want you to come to my study club to deliver a seminar on telephone communication. There is so much that can and needs to be done on the telephone to make the practice stronger. It seems to me that there is a big difference between someone getting on the phone and calling down a list of people to schedule hygiene appointments and someone making excellent telephone calls that really let people/patients know that you care and that you are concerned.

"The goal, it would seem, is the scheduling of the appointment. However, that is a great deal more difficult than it seems. There are specific skills that can and need to be learned to turn 'telemarketing' into 'tele-relations' in the dental office. Come and teach us those skills."

That was the thrust of an energetic conversation I had with Dr. Mitch Cantor, a periodontist in Southhampton, NY. Dr. Cantor understands that when a member of the team is on the telephone, he/she can make or break a relationship with a patient—and can certainly have a strong impact on whether or not a person schedules an appointment.

In this portion of the chapter on telephone communication, the critical factors of scheduling an appointment over the phone will be detailed. I will also give you armamentarium that may prove helpful in stimulating interest in the appointment, even if the patient was originally less than enthusiastic.

When you are on the telephone trying to schedule an appointment for the hygienist or for the doctor, you are, in essence, making a sales call: you are selling the benefits of the appointment. The goal of this sales call is to schedule the appointment.

Without question you believe that the scheduling of the appointment is important or you wouldn't be making the call in the first place. A strong belief in the benefits of your services give the appointment effort greater strength—strength that is driven by your sincere concern for the patient.

FOUR STAGES OF AN EFFECTIVE TELEPHONE CALL

There are four stages of any call cycle. Each type of telephone call has the same four stages. Those stages are:

1. Opening
2. Gathering information
3. Presentation
4. Closing

Let's explore those four stages. Learning how to effectively move through those four stages to reach your goal of scheduling the appointment will give you a much higher success rate.

Stage 1: The Opening

You may encounter an objection right from the beginning. Therefore, your opening is critical. If you do receive objections right from the start, you have to be poised to open the door and keep it open.

Why would you get objections at the opening of your call? Because you may be an interruption. That is one of the things that must be realized about the telephone. Some people view a telephone call as an interruption. Therefore, you must deal effectively with this.

I encourage you to use scripts for reference. Carefully planned and practiced scripts, appropriate for the different types of calls you will be making will give you the verbal skills and confidence necessary to carry out an excellent telephone call.

Don't shoot from the hip. This will get you backed into a corner more times than not. Be well prepared. The scripts will prove to be a wonderful and useful road map for you. You must then personalize the scripts to fit you and your situation. The scripts must include an opening that will stimulate the patient's interest. They must also include convincing responses not only to objections, but also to such intermediate queries, such as "tell me more." In other words, be prepared.

In the opening sequence of your telephone call you only have a few seconds (1) to qualify a person as someone who is interested in scheduling an appointment or (2) to turn an objection around.

Here is an example of an opening statement for a hygiene retention telephone call:

HYGIENE COORDINATOR: Mrs. Jones, this is Cathy with Dr. Jameson's dental office. How are you today?

Mrs. Jones, Dr. Jameson has been reviewing your records and finds that it has been seven months since your last dental cleaning and examination, and he was concerned. He asked that I call you today to schedule an appointment for your cleaning and examination. Tell me, which is better for you: morning or afternoon?

The hygiene coordinator has led into the conversation in a very affirmative, positive manner. If the patient is interested in scheduling an appointment, the coordinator has done all the right things to open that door. Notice that she did not set herself up for any no's. She controlled the entire conversation, but let the patient feel that they were in control.

However, if the patient doesn't seem overly interested in scheduling and begins to pose objections, the coordinator needs to be prepared to move the direction of the conversation into a more positive arena. She can do this by introducing the *basic turnaround*.

Let me introduce you to the basic turnaround. The basic turnaround is a way to communicate to your patients that you have heard their objection to the scheduling of the appointment and that you understand their concern. The basic turnaround allows you to continue to move in a forward direction by introducing something new to the patient. In other words, you begin to turn that objection from a negative into a positive result.

Steps of the basic turnaround:

1. Tell the person that you have heard their objection before from people who ended up going ahead with your services.
2. Introduce a new proposal.

After your opening statement—or your initial effort to schedule an appointment—patients may object. When they say they're not interested, restructure your opening and ask their permission to go on. Then introduce something new.

Gain permission to ask a couple of questions so as to evaluate where the patient is coming from. This is critical. You must qualify the person. In other words, you must know something about their concerns or their wants. Once you have this insight, you can decide which new proposal you will offer.

For example:

PATIENT: Oh, I know I need to come in, but I just don't have time right now and money is pretty tight for me here at tax time.

HYGIENE COORDINATOR: Your schedule is quite hectic, I know, and I understand the tax problem. We were hit pretty hard this year, also. Mrs. Jones, may I ask you one quick question?

(This helps the person focus on the proposal rather than on their objection.)

Once you have received their permission to ask that question, do so. Ask a question that will open a new door. Introduce something new to the patient—something beyond the original reason for the telephone call.

It could prove effective to use the feel, felt, found method here.

HYGIENE COORDINATOR: Mrs. Jones, I understand how you feel about investing in dental care at this time of the year. Many of our patients have felt the same way and have expressed a concern this year. Then they found that in our practice we have a convenient method of long-term payment that allows you to receive the care that you need without putting financial stress on you. You can finance your needed dental care and spread the payments out over a long period of time, keeping those payments small and reasonable. With this new method of payment, you can stay on a regular program of hygiene and health but not stress yourself out financially. Would this be of interest to you?

This does the following:

1. It lets patients know that you empathize with them and that you are not upset with them for their feelings or concerns.
2. You let them know that they are not the only ones in your practice who have experienced these concerns.
3. You let them know that other people have found a solution to the objection.

Prepare carefully for any sequence of telephone contacts—different telephone calls for different types of scheduling scenarios. Identify normal objections that come up on a regular basis. Write scripts that you can use to effectively deal with these objections. Integrate the basic turnaround into these scripts. This will give you flexibility and the ability to forecast problems ahead of time. It is better to be prepared than to be hit on the blindside with an objection.

Stage 2: Gathering Information

It is at this point that 75% of the actual process of scheduling presentation takes place.

The key to success here is to ask questions and to listen!

This is where the professional is distinguished from the amateur. You must:

1. Gather information from the patient's perspective
2. Know how much detail is appropriate
3. Uncover the objections
4. Figure out the patient's level of knowledge about your services

This is where you earn the right to make a presentation!

Let the patient perceive that you care. You are there to solve problems and to change smiles.

For example:

TREATMENT COORDINATOR: (Making follow-up telephone calls to

> people who have dentistry diagnosed but left untreated)
>
> Mrs. Jones, this is Pam with Dr. Jameson's dental office. I'm glad I am able to reach you.
>
> Mrs. Jones, last week during our consultation appointment you told me that you needed to speak with your husband about the treatment that Dr. Jameson had recommended to you. We were sorry that he was ill and could not come with you to the consultation. What did he think of the photographs from our imaging system that we sent home with you?
>
> Mrs. Jones, as you had requested, I am calling this week to see if you or your husband have any questions about the treatment that Dr. Jameson has recommended. I thought that I might be able to answer those questions for you.
>
> I know this is the type of treatment you would like to receive, and we want to do everything we can to make that possible for you. What questions do you have?

The coordinator had asked the patient permission to call. She had made a note to herself in her tickler system to make that call and now she is following up, just as she had said she would. She opened her conversation courteously, reminding the patient that she had requested the call. Then she very encouragingly and positively opened the door for questions. Remember, she is looking for objections. Once these objections are on the table, she has a chance to overcome each and every one. Unless she identifies those objections—by asking questions—she doesn't have a chance to overcome the objections.

Stage 3: Presentation

During this phase of the telephone contact, you must gain the person's trust and confidence and you must stimulate interest. Remember: a person will buy what they want long before they will buy what they need. Therefore, in order to get the attention you will need to schedule the appointment, you will have to key in on the person's wants: the motivational hot button.

When you present the possibility of scheduling appointments, you must stress the benefits to the patients in terms of how the treatment will meet their needs. You cannot do this unless you have asked careful questions and have determined that particular patient's needs. You must, then, be prepared to give proof that your practice and your doctor or hygienist can meet those needs. One of the most effective ways to do this is to use examples or testimonial stories of other patients.

For example:

Mrs. Jones: Yes, we looked at those pictures and my husband sees the mess in my mouth, but he isn't sure that my mouth can be fixed! And he sure doesn't want to spend the money to do so!

Treatment Coordinator: Mr. Jones isn't clear about the treatment that Dr. Jameson is recommending to restore your mouth, and he wonders if we can get results that you will be happy with, is that right?

Mrs. Jones: Yes, that's right. He thinks I'm hopeless.

Treatment Coordinator: Mrs. Jones, let me send you some other photographs of other cases similar to yours. I'm sure your husband will see the similarity. I will send the photographs that show the situation before treatment and after treatment. You saw these when you were in the office. Mr. Jones will be pleased to see that other people who have been in a critical situation such as yours have been restored. They look great, chew well, feel fabulous, are much healthier, and are a lot happier.

I will call again in a few days, if that is okay, and will try to speak with Mr. Jones to answer questions. I will also send you some information about our financial options so that when we discuss your treatment we can find a convenient way to finance that treatment. Should I call next week, or would the following week be better?

Then she would mail the appropriate material and would make a note to herself to make a phone call when the patient requested.

Stage 4: Closing

Your goal is to schedule an appointment. This is the closing. Actually, closing starts at the beginning. It is a natural part of the sequence. When all concerns and objections have been identified and a determination of how needs will be met is established—then close.

Closing means that you ask the patient for a commitment to proceed. You need to ask a patient to proceed. They are often looking for help in making that decision. Most people cannot (or choose not to) make decisions. If you do not call for a decision, you are giving your patients an excuse to procrastinate. Don't do that! Ask them to proceed. Ask them to schedule that appointment. If you sincerely believe in the dental care you are providing, who loses if the patient falls out of treatment?

For example:

TREATMENT COORDINATOR: (After she has made the next telephone call, has answered all questions of Mr. and Mrs. Jones, and has discussed the available financial options)

Mrs. Jones, do you have any further questions about the treatment or about the financial responsibility?

No?

Then let's go ahead and schedule your first appointment. Do you prefer mornings or afternoons?

HELPFUL HINTS FOR
EFFECTIVE APPOINTMENT SCHEDULING

1. Be sincere.
2. Slow your speech.
3. Use scripts.
4. Listen, listen, listen.
5. Mentally review positives and negatives and be prepared to address them.
6. Speak in layman's language.
7. Monitor and record the results of your calls.

TRACKING AND MONITORING YOUR TELEPHONE CALLS

Tracking your telephone calls and contacts is critical. It lets you analyze your own progress. This will also let you know if you are on target and are getting great results or if you need to make adjustments. If your tracking indicates that great results are not being gained, then make necessary changes. Don't keep doing something if it isn't working for you.

HELPFUL HINTS

1. Track your numbers: Don't do any telephone work without monitoring your results.
2. Set specific goals: Try to reach decision makers 50% of the time. Try to schedule one out of three of these people. Try to maintain a 50% ratio of positive results from your calls.
3. Be an ambassador for your practice: Tell everyone what you do every chance you get. Hand out those cards: Talk about your practice, your position in the practice, your doctor, and the services your are providing. Ask for referrals! 100% of all people will either need a dentist themselves or will know someone who does. Give out your cards.

Even though you may get answering machines, disconnects, and people who won't schedule, these are all stepping-stones to success. Keep trying; the numbers will be in your favor (see Fig. 9–3).

Telephone Tracking

Date	Time	# of Calls	Positive Responses	Negative Responses	Appts. Set	No Contact	Comments

Figure 9–3 Tracking telephone calls is critical so that you can determine when you are getting best results.

MIND-SET

One of the keys to great success in telephone scheduling is to keep trying—keep calling. The bigger your net, the more fish you will catch. Some people are going to slip away, but when you accept these as a part of the overall process, you can see your efforts in proper perspective. The law of averages will be in your favor if you keep trying. Don't get discouraged. Don't quit, even if you get several rejections in a row. Keep imagining that the next person will say yes.

One of the greatest assets to success in scheduling appointments over the telephone is your own initiative, your own self-motivation. Motivation is defined by Webster in the following manner: "Affecting the environment in such a way that your efforts will be more productive."

Constantly give yourself positive affirmations, such as:

1. Our services are needed by our patients.
2. I am confident.
3. This will be a great day.
4. This is *the* practice.
5. I can do whatever I decide to do.

When making telephone calls:

1. Get ready.
2. Warm up.
3. Get mentally prepared.
4. Know your product and service.
5. Describe these in patient terms.
6. Use simple, everyday language.
7. Know your practice.

Three qualities to project in your telephone calls:

1. Competency
2. Authority
3. Control

Don't take rejections personally. Use the telephone tracking device graph. See your progress. See the results of your efforts. Know that if people do not schedule appointments—and you have done the best you can possibly do—they are rejecting proposals. They are not rejecting you.

It is only through commitment to the ongoing process of personal development that you will get the results you want.

How to Handle the Irate Caller

I'm sure that by now you agree with me that telephone skills are critical to establishing and maintaining excellent relationships with patients. But sometimes that's not easy! From time to time an angry, or irate, patient calls, testing all of your telephone skills, your people skills, and your patience!

How do you effectively deal with an irate person on the phone?

1. Hear the person out.

   Don't interrupt with questions, or comebacks, or defensive responses. Passively listen. That means, encourage them to go on, to spill their guts, to get it all out!

   You passively listen by saying things, such as:

   "I see."
   "Tell me more."
   "Please, explain what you mean."
   "Oh, really."
   "I understand."
   "I agree."
   "I know."
   "You have every right to feel that way."

   Try to agree with the caller where possible.

2. Actively listen.

   The single most effective skill you can use to defuse anger and to calm the irate person is to listen. Encourage a person to tell you about his/her concern. Reflect back to the person (in your own words) what you think you hear. This will give you the opportunity to clarify. This will let the person know that you are trying to understand. This will also calm the person down.

3. Do *not* argue with the person.

This will only add fuel to the fire. Your defensiveness will add to their defensiveness. The emotionality of the conversation will become so intense that the possibility of moving into a problem-solving mode may be nullified.

4. Control your urge to become angry!

A person can't stay angry with you if you remain calm. Your calmness will let them know that you respect them, that you *are* concerned about their problem, and that you want to hear all about the problem so you can do something about it. If a person senses your willingness to listen and to work on the resolution of the problem, he/she will be more inclined to work with you in return.

Once you've heard a person out, you've passively and actively listened, you've defused the anger by not being negatively stimulated, say thank you to the person for letting you know about the problem—and *mean it!*

"Thank you for calling, Mr. Johnson. I am so glad to know that you have a concern. I really want to help you. I appreciate your honesty."

5. Ask Mr. Johnson to repeat his concerns so that you can write them down.

"Mr. Johnson, would you please repeat your concerns for me? I am going to write these down so that I can see if I can help you and so that I can make sure I accurately inform Dr. Jameson of your concerns."

6. Once you have recorded the necessary information, repeat it to him to determine your correct interpretation.

7. Let him know that you understand his concerns and that you will do your best to find appropriate answers. Say you will get back to him, and then *do so!*

TEAM MEMBER: Mr. Johnson, I understand your confusion over this situation. Now that I have all of the correct information, I am going to discuss this with Dr. Jameson and I promise we will find an answer to your questions. I will get back to you within the next couple of days. Would you prefer that I call you at work, or would it be better if I call you at home?

Do what you say you will do. Find the answers. Get a solution. Make a note in your tickler system to get back to him in the next couple of days at whatever number he gave you and then—*do it!*

GREAT RESULTS

If you will commit to going through these steps, you will find that most irate callers will calm down. Oftentimes, by the second time they have repeated their concerns, they will have either (1) softened or (2) answered their own questions.

People will reflect your temperament. If you become angry, they will become angrier. If you slow your speech, lower your voice, remain calm, they will reflect this. Only when the irate caller is calmed can a solution be generated.

A side benefit to this process is that you will not be so stressed out. If you become angry, your blood pressure will rise, your pulse will increase, and stress will negatively affect your day and your performance. However, if you choose to be the one in control, you will encourage a calmer patient and you will control your own stress.

IN SUMMARY

It has been said that the telephone may be the most important instrument in your office. As such, the value of excellent telephone technique cannot be overstressed. The bottom line is this: you have about 30 seconds to educate, to motivate a person to come to your practice, to let a person know you understand their concerns, or to let a person know you are ready to assist them.

Be informed about the policies and procedures of your office. A person makes a judgment about the organization and services available in your office by the person who answers the telephone. Therefore, anyone—everyone—who answers the phone or speaks to any clients/patients on the phone for any reason needs to be adept at telephone skills.

We talk a lot about marketing tools in dentistry. One of the most valuable marketing tools in your office is the telephone. Use it carefully and powerfully.

Time spent on studying, reviewing, and practicing excellent telephone technique will enhance your practice-building opportunities.

Great Telephone Communication = Great Production Opportunity.

CHAPTER 10

MAKING AN EFFECTIVE CASE PRESENTATION: GAINING TREATMENT ACCEPTANCE

"People don't care how much you know until they know how much you care—about them."

ZIG ZIGLAR

In the majority of today's dental practices, fabulous opportunities lie *within* the walls of those practices. Most dental practices can double the amount of dentistry presently being provided by nurturing that which they already have: their existing patient family. Excellent communication is the bottom line to great case presentations, which lead to treatment acceptance.

For some dental professionals, this is where the "rubber meets the road." They know that case presentation makes the ultimate difference, but are extremely uncomfortable—or inadequate—with the communicative skills required to make a great case presentation. In addition, some teams aren't where they want to be with their practices, but are not even aware of that fact that the presentation scenario is weak. They just go along telling people what they need, explaining everything very technically, and then wake up in the middle of the night thinking, "I wonder if Mr. Jones ever scheduled an appointment." Stressful!

The *most* crucial marketing strategy that a practice embraces, the one where practice growth and stability begins and ends, is a well-managed practice. Perhaps the most critical of these management strategies is the system of case acceptance. Case acceptance is the fulcrum of your practice: diagnosis, treatment planning, case presentation, and follow-up. Knowing how to do the dentistry is essential. However, knowing how to do it and getting to do it are sometimes two separate things!

Most practices can double from within! As I said earlier, most practices have more dentistry sitting in the charts waiting to be done than they have ever provided. Getting the dentistry out of the charts and into the mouths is critical for practice growth. In addition, it costs less to nurture an existing client than it does to find a new one.

So combine those two pieces of information. Develop a business plan that accomplishes the following: (1) doubles the practice by nurturing your own patient family, and (2) does this in a cost effective manner.

TREATMENT ACCEPTANCE: THE GOALS

Developing an effective protocol for your treatment presentations accomplishes the following goals:

1. Patients outline the goals they want to accomplish.
2. Patients are educated about the need for and the benefits of the dental treatments that you are recommending.
3. Patients become motivated to accept those particular recommendations.

The purpose of a case presentation protocol is to encourage patients to say yes to the treatment you are recommending. I believe so much in the dentistry that John is providing that I hurt when a patient walks out the door not scheduling an appointment to proceed. I think that it is our responsibility to present the recommendations so excellently that the person will have every chance in the world to go ahead. If we don't do a good job of presenting the case and the person does not schedule, I think the patient loses. Therefore, learning the sophisticated skills of case presentation is critical for the success of the practice and for the benefit of our patients.

The Team Approach to Case Acceptance

Everyone on the dental team has specific responsibilities. The case acceptance protocol is not just one person's responsibility. It is not the responsibility of the doctor only. Everyone is critical. Everyone has dynamic responsibilities. Know that each person on the team has what business experts call a "moment of truth." Each person on the team can make or break a person's willingness to proceed with treatment.

I have spent a couple of decades studying with some of the great business consultants in America. Why? To learn what skills are essential for corporate success. I have listened to them with my "dental ears" and have tried to translate the information to our own industry. The following steps of case presentation are patterned from the format promoted by the great companies of the world. They know that if they want their products or services to be purchased by their consumers/clients that they have to study and work on presentation skills. So do we, as dental professionals. Here, indeed, is the moment of truth: the case presentation. Study. Practice. Make a commitment to learn these intricate skills so that they are as comfortable for you as that crown prep!

This chapter on case presentation pulls together the communication skills studied in previous sections. Here you will see that the foundational communication skills become strong vehicles throughout the presentations. You must have the strong foundation before climbing to a higher level of expertise.

STEP 1: BUILD THE RELATIONSHIP

Before a person will purchase your service, you must first build a relationship of confidence and of trust. Without this level of confidence or trust, you will not get to the point where you can provide necessary treatment. Remember that the oral cavity is an intimate zone of the human body and deserves the ultimate respect. This is why establishing trust is so critical. If you are going to restore a mouth to health again or if you are going to change a person's smile, that person must have ultimate trust of you.

Most of the time, the initial contact is made on the telephone. The telephone may be the single most powerful marketing tool in your practice (see Chapter 9, Communication By Telephone). People calling your dental office make a subconscious decision about the dental treatment they will receive by the treatment they receive on the telephone. In fact, they often make a decision about whether they will come to you or go to someone else based on their first impression on the telephone. It is very important that the person answering the telephone be enthusiastic, warm, and knowledgeable. They must concentrate when answering that telephone and focus on the business at hand, which is the person on the other end of the line.

The New Patient

Upon answering the telephone, if you determine this is a new patient calling, immediately grab a patient communication slip (see Fig. 10–1) and begin recording information about this newcomer. You begin gathering information so that you can be totally prepared for the patient's arrival and so that you will know something about him/her before the appointment.

Welcome Packet

Send a packet of information to new patients before their scheduled appointment. Why? To prepare for their arrival and to get them acquainted with you and your practice. Begin the bonding process in advance. This will offset some of those new-patient broken appointments and no-shows!

In this Welcome Packet include the following:

Patient Communication Slip

Date: ______________________ New Patient ☐ Patient Record ☐ Appt Date: ______________

Name: ______________________(Pronounced) ______________________ Adult ☐ Child ☐ Age of Child _____

Referred by: ______________________ Family Member ☐ Friend ☐ Parent/Guardian of Child: ______________

Purpose of Call: Emergency ☐ Examination ☐ Other ______________________

Symptoms: ______________________

____

____

____

Date of FMX: ______________________Doctor's Name: ______________________

Premedication Necessary: ______________________

Insurance Information:

Employer: ______________________Insurance Carrier: ______________________

Comments: ______________________

____

____

____

Address: ______________________

Street City State Zip

Home Phone: ______________________Work Phone: ______________________

Figure 10–1 The telephone communication slip lets you gather and store essential information on new patients and on emergency patients.

1. Patient Information Sheet/Health History
2. Confirmation card for the appointment, such as the one in Fig. 10–2.
3. Practice Brochure or Welcome Letter
4. Patient Education Newsletter
5. Self-Addressed Stamped Envelope

Tell patients that you will be sending them some information about the office. Ask that they complete the patient information sheet/health history and mail it back to you in the enclosed, self-addressed envelope. Tell them that by returning this information prior to their appointment that you can be better prepared for their visit and that

WELCOME CARD

WELCOME
TO OUR OFFICE!

We're looking forward to meeting you!
Because you are a special person, we
will strive to make your visit with
us a unique and pleasant experience.

This time has been reserved especially for you...

Date

JOHN H. JAMESON, DDS
101 Jameson Drive · Wynnewood, OK 73098
Phone: (405) 665-2041
©Jameson Management Group

Figure 10–2 A card confirming the appointment of a new patient helps them to feel special, alleviating some no-shows.

you will be able to seat them quicker. Sell the benefits of this request to patients and they *will* respond positively by sending the information back to you.

For example: "Mrs. Jones, I am going to be sending you some information about our practice. We want you to know about us before your visit. Included in your packet will be a brochure about our office, one of our newsletters, a card confirming your appointment, and your information sheet/health history. We ask that you complete this information sheet and send it back to us in the self-addressed envelope that I am including. Most of our patients are much more comfortable completing this information sheet at home where they have all of the necessary data. And by doing this, we can have all of your information in our computer, be totally prepared for your visit, and be able to seat you much quicker. "

In this conversation with the new patient, the request is presented in terms of how her compliance will benefit *her*—convenience, preparedness, and quicker seating. All of these issues are benefits to the patient. Know that you will get much farther with *any* request if you present that request in terms of how it benefits the other person. Remember: People want to know, "What's in this for me. How will this benefit me?"

Practice Brochure

By enclosing a brochure about your office (see Fig. 10–3) that outlines the positive aspects of your practice, you can offset last minute cold feet and, in addition, you can build a person's confidence in you before they step a foot inside your door. Don't just list the services you provide. Think of the brochure as a marketing piece for and about your practice.

A practice brochure needs the following components:

1. Practice logo.
2. Color scheme that is consistent with all other written communication pieces.
3. Mission statement.
4. Sell the benefits of the services you offer. Don't make it a nagging piece about things they have to do when they come to your office.

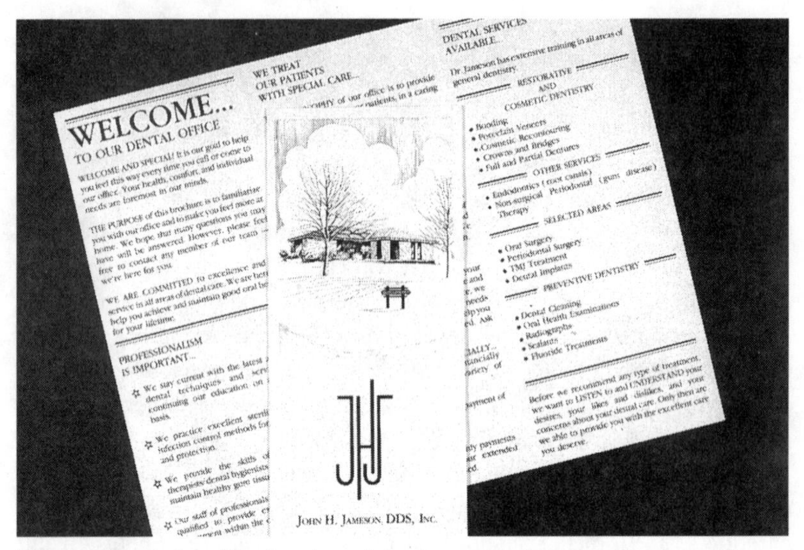

Figure 10–3 An office brochure introduces you to new patients and is a first step in establishing a bond.

5. Open your own doors for the types of services you wish to develop in your practice.
6. Photograph and bio of the doctor/doctors to build confidence in their skills.
7. Directions to the office (map, if appropriate).
8. Telephone numbers.
9. Office hours.
10. Attractive, to the point, encouraging, motivational, classy.

Patient Education Newletter

A patient education newsletter (see Fig. 10–4) sets you apart from the average. It lets people know something about the services you make available, and it lets people know of your commitment to total, long-term care.

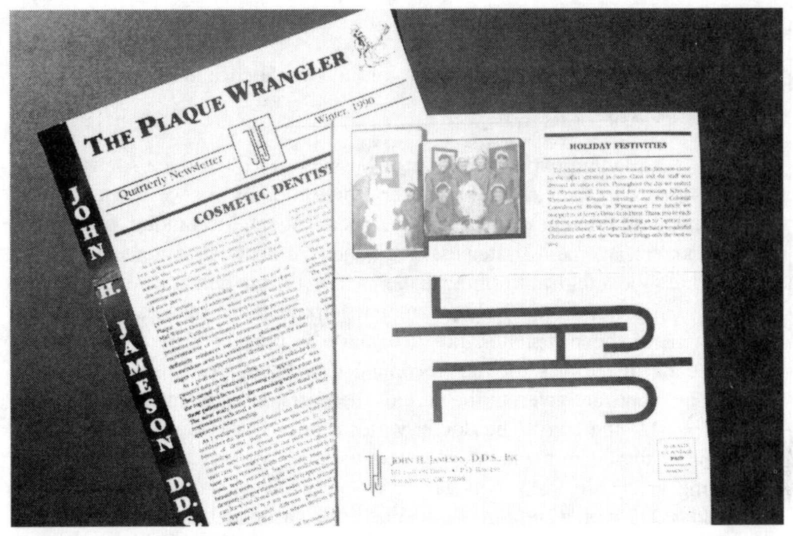

Figure 10–4 Patient education newsletters become excellent vehicles for continuous communication with your patients.

PATIENT EDUCATION NEWSLETTERS

Surveys indicate that patients do like a personally produced newsletter. The newsletter should be brief and personal; patients should sense that they are in direct contact with you, that you are sharing something special with them, that you care about them, individually.

We have produced a patient education newsletter in our own practice for the past 10 years. We try to accomplish the following goals with our newsletter:

1. Be in the homes of each of our patient families in a positive way on a regular basis
2. Inform and educate patients about what's happening in dentistry
3. Let people know what we are doing in our own practice to stay on top of the latest and greatest in dentistry

4. Express our appreciation for their confidence in us

5. Ask for referrals

Marketing experts encourage businesses to introduce something new to their client base every three months. So we discuss something different every quarter: a specific dental technique, one of our management strategies, or a beneficial health-care note. We address the total health of our patients, not just the dental health.

All members of the team participate. One person is in charge of the newsletter, but delegation is practiced. At a team meeting, the topics for the upcoming newsletter are determined. Assignments are given to the various members of the team who are to provide the copy. The doctor writes a brief article for the cover page. Other team members volunteer to either write an article or to bring an article from a magazine or news piece. If an article is under 200 words, reprint is acceptable. If an article is more than 200 words, permission needs to be accessed and credit needs to be given.

By a certain date, everyone has their data to the person in charge. She puts the newsletter together. Presently, we produce a one-page newsletter (front and back), legal size. Short, sweet, informative, comfortable, attractive.

Our local printer has our logo on file. He changes the date appropriately. Once the original format was created, it has been a snap from that point forward. The printer takes our copy and gets it ready for our final proof. He prints and folds the newsletter. He places our return address, the bulk mailing number, and address correction notice on the outside fold.

Our computer produces the mailing labels, but professional companies can do this for you if you are not computerized. At one staff meeting each quarter, the entire team gathers together over lunch (doctor buys!) and we label the newsletters and place them into zip code order, a requirement if you are using a bulk mailing number.

We invest approximately $600 and about six total hours (everyone combined) per quarter. We feel that the time and money are well worth the effort. From one newsletter we generate new patients and a great deal of dentistry. But we never lose sight of our number one goal: to educate our patients about the benefits of the dental services we provide.

Face-Face Contact

Building the relationship continues as the patient is received in the office. The person greeting the patient needs to stop what he/she is doing, stand up, and make a conscious effort to greet the patient by name. An introduction is desirable.

For example:

> TEAM MEMBER: Mrs. Jones? I'm Cathy. I spoke with you on the telephone. Welcome to our practice. We're glad you are here. Thank you for sending your information back to me. We have all of the necessary information, and so the doctor will be right with you. Make yourself comfortable for just a moment, and I will let the team know that you are here. By the way, I notice that John Smith referred you to our practice. He's great. We really appreciate him telling you about us.

Usually the next person that a patient meets is the clinical assistant. I encourage the clinical assistant to address the person in this way:

> ASSISTANT: Mrs. Jones, I'm Jan. I'm Dr. Jameson's clinical assistant, and I will be working with you today. You may come with me.

Then the clinical assistant escorts the person to the clinical area, or in some offices, to the consultation area. The assistant spends a few minutes in personal conversation, then reviews the health history: making sure that it is completed, asking some pertinent questions, letting the person know that she wants to get to know her as an individual, and that she notices *anything* about her that might affect her treatment. Tell the patient a bit about the philosophy of the practice and inform the patient as to what will be happening today. It is very important for a person to know what is going to happen before it happens. The clinical assistant can go through the scenario of the initial examination, making sure the person is informed and comfortable.

Upon the entrance of the doctor, the clinical assistant can introduce the doctor. Apply the same etiquette that you use in your homes to the dental office. Doctor Jameson walks in and Jan says, "Mrs. Jones, this is Doctor Jameson. Doctor Jameson, I'd like to introduce you to meet Mary Jones."

The clinical assistant could then give the doctor a bit of appropriate information about Mrs. Jones so the doctor can comfortably begin the conversation. The referral source is a good starting point.

While the patient is still sitting up, the doctor sits down, moves his chair around so that eye contact is established with the patient and an adult-to-adult body posture is maintained. Following appropriate social graces, the doctor moves into Step 2 of the treatment acceptance scenario.

STEP 2: ESTABLISH THE NEED

Your patient information sheets should have questions that the person can answer about their attitude toward their dental health or their attitude toward the appearance of their smile. In addition, patient questionnaires or surveys can give you valuable information about your entire patient family and, certainly, about individual concerns.

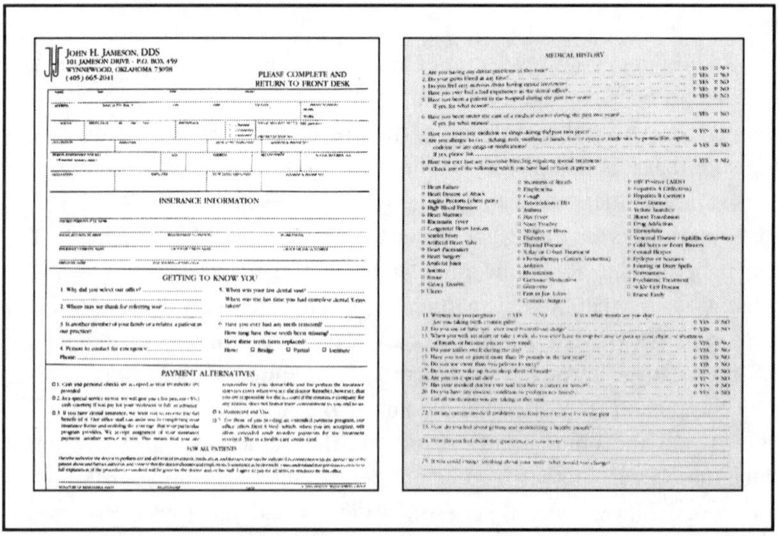

JOHN H. JAMESON, DDS
101 JAMESON DRIVE · P.O. BOX 459
WYNNEWOOD, OKLAHOMA 73098
(405) 665-2041

PLEASE COMPLETE AND RETURN TO FRONT DESK

[illegible]

INSURANCE INFORMATION

[illegible]

GETTING TO KNOW YOU

1. Why did you select our office?
2. Whom may we thank for referring you?
3. Is another member of your family or a relative a patient in our practice?
4. Person to contact for emergency
Phone:
5. When was your last dental visit?
When was the last time you had complete dental X-rays taken?
6. Have you ever had any teeth removed?
How long have these teeth been missing?
Have these teeth been replaced?
How: ☐ Bridge ☐ Partial ☐ Denture

PAYMENT ALTERNATIVES

☐ 1. Cash and personal checks are accepted as treatments are provided.
☐ 2. As a special service to you, we will give you a five percent (5%) cash courtesy if you pay for your treatment in full in advance.
☐ 3. If you have dental insurance, we want you to receive the full benefit of it. Our office staff can assist you in completing your insurance forms and [illegible] the coverage that your particular program provides. We accept assignment of your insurance payment [illegible]. This means that you are responsible for your deductible and the portion the insurance does not cover [illegible]
☐ 4. Mastercard and Visa
☐ 5. For those of you needing an extended payment program, our office offers [illegible] which, when you are accepted, will allow extended [illegible] payments for the treatment received. This is a health care credit card.

FOR ALL PATIENTS

[illegible]

MEDICAL HISTORY

1. Are you having any dental problems at this time? ☐ YES ☐ NO
2. Do your gums bleed at any time? ☐ YES ☐ NO
3. Do you feel very nervous about having dental treatment? ☐ YES ☐ NO
4. Have you ever had a bad experience in the dental office? ☐ YES ☐ NO
5. Have you been a patient in the hospital during the past two years? ☐ YES ☐ NO
If yes, for what reason?
6. Have you been under the care of a medical doctor during the past two years? ☐ YES ☐ NO
If yes, for what reason?
7. Have you taken any medicine or drugs during the past two years? ☐ YES ☐ NO
8. Are you allergic to [illegible] penicillin, aspirin, codeine, or any drugs or medications? ☐ YES ☐ NO
If yes, please list
9. Have you ever had any excessive bleeding requiring special treatment? ☐ YES ☐ NO
10. Check any of the following which you have had or have at present:

☐ Heart Failure
☐ Heart Disease or Attack
☐ Angina Pectoris (chest pain)
☐ High Blood Pressure
☐ Heart Murmur
☐ Rheumatic Fever
☐ Congenital Heart Lesions
☐ Scarlet Fever
☐ Artificial Heart Valve
☐ Heart Pacemaker
☐ Heart Surgery
☐ Artificial Joint
☐ Anemia
☐ Stroke
☐ Kidney Trouble
☐ Ulcers
☐ Shortness of Breath
☐ Emphysema
☐ Cough
☐ Tuberculosis (TB)
☐ Asthma
☐ Hay Fever
☐ Sinus Trouble
☐ Allergies or Hives
☐ Diabetes
☐ Thyroid Disease
☐ X-Ray or Cobalt Treatment
☐ Chemotherapy (Cancer, Leukemia)
☐ Arthritis
☐ Rheumatism
☐ Cortisone Medication
☐ Glaucoma
☐ Pain in Jaw Joints
☐ Cosmetic Surgery
☐ HIV Positive (AIDS)
☐ Hepatitis A (Infectious)
☐ Hepatitis B (Serum)
☐ Liver Disease
☐ Yellow Jaundice
☐ Blood Transfusion
☐ Drug Addiction
☐ Hemophilia
☐ Venereal Disease (Syphilis, Gonorrhea)
☐ Cold Sores or Fever Blisters
☐ Genital Herpes
☐ Epilepsy or Seizures
☐ Fainting or Dizzy Spells
☐ Nervousness
☐ Psychiatric Treatment
☐ Sickle Cell Disease
☐ Bruise Easily

11. Women: Are you pregnant? ☐ YES ☐ NO If so, what month are you due?
Are you taking birth control pills? ☐ YES ☐ NO
12. Do you use, or have you ever used recreational drugs? ☐ YES ☐ NO
13. When you walk up stairs or take a walk, do you ever have to stop because of pain in your chest, or shortness of breath, or because you are very tired? ☐ YES ☐ NO
14. Do your ankles swell during the day? ☐ YES ☐ NO
15. Have you lost or gained more than 10 pounds in the last year? ☐ YES ☐ NO
16. Do you use more than two pillows to sleep? ☐ YES ☐ NO
17. Do you ever wake up from sleep short of breath? ☐ YES ☐ NO
18. Are you on a special diet? ☐ YES ☐ NO
19. Has your medical doctor ever said you have a cancer or tumor? ☐ YES ☐ NO
20. Do you have any disease, condition or problem not listed? ☐ YES ☐ NO
21. List all medications you are taking at this time
22. List any current medical problems you have been treated for in the past
23. How do you feel about getting and maintaining a healthy mouth?
24. How do you feel about the appearance of your teeth?
25. If you could change anything about your smile, what would you change?

Figure 10–5 Health history can serve a dual purpose: (1) a way to gather necessary clinical information and (2) a way to dertermine a patient's main motivation.

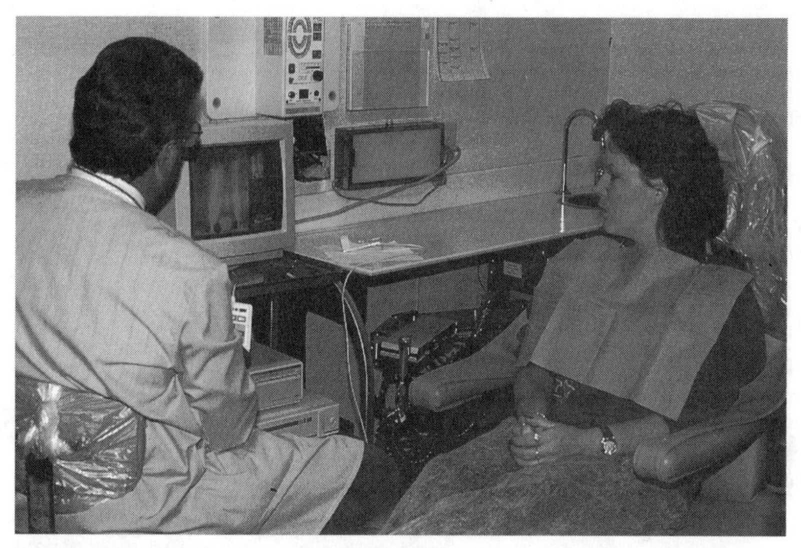

Figure 10–6 Digital radiography illustrates areas of need for a patient. Superior education!

Figure 10–7 An intra-oral camera is the single best tool available for patient education.

PATIENT SURVEYS OR QUESTIONNAIRES

Successful corporations do surveys and questionnaires on a regular basis so that they change with the times and with the changes in client desires. The companies that listen to their clients and respond rather than react are usually one step ahead of the competition.

Gathering information from your patients puts you in a position to respond to the wants and needs of your patients. You can find out what they like about you so that you can do more of that. You can find out what they don't like so that you can consider change. You can also determine what services might be of interest to your patients so that you can begin to offer those services.

The survey should be short and easy to complete. The questionnaire can be structured for yes/no answers, or you could structure the survey in an evaluative style, such as, "Rate the following from poor to excellent." Then list 10–20 questions for the rating. Close the survey with an open-ended question, such as "If you could change one thing about our practice, what would you change?" or "What would you like us to offer that we are not presently offering?" or "If you were going to recommend us to a friend or family member, what would you say is our best quality?"

You can either send the surveys out to your patient family, or you can ask them to complete the survey while they are in the office. If you mail the surveys, make some kind of special offer if they return it to your office. Be aware that if people are going to offer constructive criticism of your practice that they may not want to sign the survey. That's okay.

As you are asking people to complete the survey, explain that you are trying to improve your services and want their input so that you can be sure you take excellent care of them. Presented in this manner, most people will respond positively and will complete the survey for you.

Patient surveys are a fabulous way to find the strengths and weaknesses of your practice. What a powerful way to evaluate your position and to set goals for improvement!

If you do ask written questions of your patients, be sure to respond to their answers. This will open doors for possible treatment modalities. It will also let the patients know that you are sincerely interested in meeting their needs. Giving the patient a sense of control over their situation is critical for empowerment. By asking questions and listening, you gain valuable information and insight. You gain control of a conversation by asking questions, but at the same time, you give the patient that needed *sense of control.*

Opening questions that get this desired result might be:

1. "Mrs. Jones, how can we help you?"
2. "Mrs. Jones, tell me, what are your goals for your teeth, your mouth, and your smile?"
3. "Mr. Smith, tell me what you like most about your teeth. Then tell me what you like least about your teeth."
4. "Mr. Smith, if there were anything you could change about your smile, how would you change it?"

Once you ask a question it is very important that you stop and listen. Reflect back to the person what you think you hear them saying, making sure you are hearing them clearly and accurately. (Actively listen.) Do not put your own value system on them.

Hank

I have the privilege of consulting for a fabulous doctor who is presently the president of his state dental association. During my first day of consulting, I always do an analysis of the practice and evaluate each and every system as it is presently being administered. One of those systems is the initial patient interview and examination.

This wonderful doctor performed one of the best initial interviews that I have ever observed. His patient was a 68-year-old woman. Her name was Hank. She was a dapper woman, fashionably dressed, hair combed nicely, well groomed, and quite friendly.

During his initial interview, the doctor used great body language. He was seated level with the patient, he leaned slightly forward, he held steady but comfortable eye contact with Hank during the entire interview. He told Hank that his clinical assistant was going to join them for the interview because he wanted her to take careful notes so that he could give Hank his full attention. (Great!)

Once the social graces were completed, this doctor asked one of the best questions I have ever heard a doctor ask. He said, "Hank, tell me, what are your goals for your teeth, for your mouth, and for your smile?" (I thought I would fall on the floor with that one! What a door opener!)

Here is what Hank said. (Notice the motivators.)

"Well, Doctor, in two months I am going to my 50th high school reunion, and I want to look great. The teeth in the front of my mouth are all a different color and I hate them. I want you to make them the same color. And on the bottom, I have "gaposis" on both sides. I have one of those things that comes in and out, but I hate it. It hurts. I didn't even bring it today, because I just can't wear it. Do you do implants? And, Doctor, my husband died four years ago and he went to his grave without his teeth. I don't want that to happen to me."

Hank's motivators were numerous.

1. Time: she had two months.

2. Appearance: she didn't like the different colors on her front teeth. She wanted them to all be the same color. She wanted to look great!

3. Comfort: her partial hurt her so much that she couldn't wear it.

4. Function: she couldn't wear the partial, so obviously her chewing was being affected. She wanted something more permanent. She inquired about implants.

5. Keeping her teeth for a lifetime: she didn't want the same dental fate that her husband had endured.

The only motivator that Hank did not mention was money. In fact, she didn't care about money. Money was not an issue for her. The power of the other motivators outweighed the issue of money. Remember: People will buy what they want long before they will buy what they need.

Then the doctor recapped what Hank had said. "Let's see. Hank, you have a special event coming up, and you want us to see about getting your front teeth to match so that you are happier with your smile. You want us to see about doing something on the bottom so that you are more comfortable and so that you can chew better. And you want us to get you healthy and keep you healthy so that you keep your teeth. Did I hear you right?"

Hank replies, "Yes, that's it."

Then the doctor asked the second best question that I have ever heard a doctor ask. He asked, "Hank, tell me, what are your expectations of me?"

Now I am on the floor! What power each of those questions created! The first question opened the door for Hank to tell the doctor her needs as she saw them. She defined her own motivators so that the doctor could be about the business of meeting those needs. And the second question opened the door for Hank to give him permission to do the very best he could to reach those goals.

If you don't learn anything else from this book, learn the value of those two opening questions in your initial interview. By empowering the patient, she is telling you what to do rather than you telling her. A person cannot be pushed into making a decision, but you can lead them into making a decision by asking the right questions. In addition, when patients tell you what they want, and you respond to those wants, you are doing just what they want. If you tell them they need something, they may doubt it. But if they ask for it, they will buy into it.

I didn't know the end of the story for a few months. When I was back in the office doing my next consult, I asked, "What ever happened to Hank? Did you get her fixed up?"

"Yes, we did veneers on the top, did a new partial for her, and have her on an excellent program of maintenance with Rhonda (the hygienist)."

"Did you get it done in two months, before her reunion?"

"Oh, yes. We became good friends with Hank. As she was getting ready for the reunion she would ask us what she should wear, what color would look best with her new smile, and so on. See, Hank knew that when she was going to her reunion, she was going to see her high school sweetheart. They became reacquainted, fell in love again, and got married. They kept both of their homes and go back and forth between the two states and are having a ball."

What a fun dental love story or human interest story. You never know. And it all began with those awesome opening questions that opened the door for Hank to express her wants and needs. The patient received what she wanted, and the doctor was able to provide some great dental care and to help a "woman in need!"

Treatment Planning

Once the initial interview is completed and you are clear about the patient's goals, begin the clinical examination. The information gathered at the comprehensive examination becomes the worksheet for your treatment plan.

At the conclusion of this initial appointment, you will have accomplished the following:

1. Established the patient's perceived need and determined their own personal goals
2. Established the clinical need through careful and comprehensive diagnosis
3. Laid the groundwork for the treatment plan

After the necessary data and information has been gathered and the comprehensive examination has been concluded, invite the patient back to your practice for a consultation appointment. Find out who the decision maker is, and invite that person to the consultation appointment, also. Tom Hopkins teaches, "There is no reason to call for a decision, if the decision maker is not there."

For example:

DOCTOR: Mrs. Smith, I need to evaluate the information I have gathered today so that I can develop a treatment plan specifically to meet your needs. I'd like to invite you back to the practice in about a week for a consultation appointment so that we can sit down together one-on-one and discuss your particular situation and my treatment recommendations. Is there anyone besides yourself who will be involved in deciding whether or not you will proceed with treatment?

PATIENT: Well, yes. My husband and I make our decisions together.

DOCTOR: Your husband? Great. Then let's schedule a consultation appointment that will work for the both of you. I think it is very important that he hear the recommendations that I will be making for your treatment. Would that be acceptable to you?

As you are scheduling these consultation/education appointments, try to schedule them within a week of the initial examination. As Tom

Hopkins says, "The time to sell is when you have a willing buyer." So don't let too much time pass before you discuss the treatment to be rendered.

STEP 3: INSTILL THE DESIRE

Here you will be educating and motivating the patient to accept the treatment that you are recommending. As dental care providers, you are *educators* of dentistry. People don't come to the dental office with very much dental knowledge. In fact, according to the American Dental Association, the number one reason people don't come to the dentist or do not say yes to the recommended treatment is "no perceived need or lack of dental education." Thus education is your biggest commission.

As educators, you must access the best methods of teaching. Learning takes place visually, then it makes sense to have excellent visual aids to show a person the end results and the benefits of treatment that you are recommending.

You want patients to know the following things:

1. What do they have now?
2. What treatment do you recommend to restore their mouth to health again or to create that beautiful smile?
3. What are the advantages or benefits of proceeding with treatment?
4. What are the disadvantages of not proceeding with treatment?
5. What is the financial responsibility?

Visual aids are mandatory if you want to get those messages across:

1. Brochures
2. Books
3. Before and after photographs
4. Educational videos
5. Models

6. Intra-oral cameras with recorded images of other similar situations

Five Ways to Turn Brochures into Educational Tools

Are your brochures hanging in a display on your wall in your reception area? Do patients read them often? Are they able to apply the information in the brochures to their own situation? Are your brochures powerful learning tools, or could they be doing more for you?

Educational brochures—when used properly—can be tremendous visual aids. They can be used to introduce dental concepts to your patients, to help you present your recommendations during the consultation appointment, and become great marketing tools.

Here are five suggestions of ways to turn your brochures into true teaching/visual aids:

1. Treatment Recommendations:

Visual aid presentations are most beneficial when certain methods of teaching are applied to their use. When you are discussing or recommending a particular treatment to a patient, pick up a relevant brochure and do the following:

- Select a brochure that shows a patient the end results of the treatment you are recommending.
- Open the brochure and with a pen or with a highlighter, accent the areas that are particularly relevant to that patient.
- Verbally inform the patient about how the brochure pertains to their situation. You might say something like:

  "Sarah, we have discussed your concern about the staining on your teeth. I have recommended a procedure called porcelain veneers. This brochure gives an explanation about this procedure—what is involved in the treatment and how the veneers look when treatment is completed. On this page you see that very little tooth preparation is necessary (highlight this) and that very little total time is involved in this treatment (highlight this). In two visits we can remove those stains and produce that gorgeous smile you want!"

- Close the brochure and write the person's name on the front and hand it to them. Then, continue:

  "Sarah, this brochure is for you. You can take it home so that you can reread it. I want you to fully understand the procedure and to know what's going to happen and what the end results of the treatment will mean for you. This brochure should answer most of the questions that you or your husband may have about porcelain veneers. But, please feel free to call about anything. We would love to help you get that new smile."

  Instructing in this manner makes each person feel special. Handled in this manner, patients will be more likely to keep a brochure, read it again, and refer to it. This also gives them the opportunity to show a spouse or other family members the dentistry you have recommended. Decisions are often made jointly. Therefore, you benefit by extending your education to your patients' families as well.

2. Newsletters:

   In your patient education newsletter, include a colorful brochure about a particular procedure you are addressing in that issue. Example: The ADA brochure on periodontal disease, entitled "Gum Disease: The Eight Danger Signs," is a fantastic brochure. As you are educating your patients about periodontal disease, you will want to include articles about the subject in your newsletter, but the articles will be greatly enhanced and will have a far greater impact if a color brochure is included in that issue.

3. Informational Letters/Mailings:

   From time to time you will want to mail special letters to your patient family telling them about a new procedure you are offering, informing them of a change in policy, making a special request of them, etc. You may have recently attended a course on a subject and want to inform and excite them about the new possibilities. In these letters, provide an informative brochure, which will provide further information and will serve as a visual aid.

4. Reception Area:

   It's fine to place or to display educational brochures in your reception area. Some people sitting in the reception area will pick up the brochures and read them. However, to encourage more reading of this literature, invite the patients to do just that! Don't assume that people will know that the brochures are theirs for the taking. Ask them to do so!

5. Continuous Care Notices:

So many hygiene/continuous care notices carry a negative message such as: "You haven't been in to see us in _____ months! Where have you been?" and so on. Instead of a negative, naggy type of reminder or notification of their delinquent appointment, why not give them a wonderful reason to come in to the office! Invite these patients to come to the practice for their continuous care appointment and also to learn about the new techniques available in dentistry today. An educational brochure, included in these invitations, can stimulate curiosity and interest in new developments—such as porcelain veneers or bleaching. This type of invitation and brochure can motivate a person to phone for an appointment and can stimulate questions at the time of the appointment. Those people who are sitting on the fence may now have a reason to come in.

Brochures can be used effectively to promote your practice and to further educate people about the options available in dentistry today. Become a teacher of dentistry. Patients will gain from the new learning and you will be able to provide more care. Increased education will lead to increased production.

Get in Focus with the Patient

At your initial interview, you established the person's emotional hot button or main motivator, whether appearance, comfort, function, keeping the teeth for a lifetime, time, or money. Knowing this vital information lets you gear your presentation toward the patient's perceived need.

As I said before, people will buy what they want long before they will buy what they need. So get on their side. You will get a great deal further with treatment acceptance if both you and the patient are after the same thing. People want to know, "How is this going to affect me?

What is in this for me? How is this going to affect my health, my looks, my pocketbook, my schedule?" Determine their main motivators. Direct your comments and your presentation accordingly.

For example:

DOCTOR: Mrs. Smith, tell me, if there were anything that you could change about your smile, what would that be?

PATIENT: Well, I just hate my teeth. They have stains all over them, and I'm embarrassed to open my mouth. I usually just cover my mouth when I smile or laugh, and I am really ready to change that.

DOCTOR: You're unhappy with the discoloration of your teeth, and you are interested in getting a whiter, brighter smile. Is that right?

PATIENT: Yes!

You know that this person's main motivational factor is appearance. When she comes back to the office for consultation, have your visual aids: appropriate books, brochures, or images on your intra-oral camera readily available. Your presentation needs to be simple and direct and geared toward the answering of that patient's perceived need.

For example:

DOCTOR: Mrs. Smith, last week when you were here for your initial examination, you indicated a concern about the color of your teeth. Let me show you an image of a person who had a situation similar to yours. She had stained teeth also. Can you see the similarity?

PATIENT: Yes. That looks just like my teeth.

DOCTOR: Once we completed her treatment she looked like this (switch to the after). What do you think?

PATIENT: That's great.

DOCTOR: Well, Mrs. Smith, I am comfortable telling you that once we complete your treatment, your smile will look similar to this. Would this be something that would be of interest to you?

And she will probably say yes. Then go into some detail about the treatment recommendation, but don't go into incredible detail on the technique itself. Oftentimes, dental professionals get so involved with the technique that they totally lose the patients. They forget that patients haven't had dental training and that you are probably speaking Greek to them. If you get too technical or use dental words, you risk confusing the person. A confused person cannot make a decision. Keep the technical aspects simple. Let the person know what the end results and the benefits are going to be. That is really what matters to them.

Keep the presentation short! No more than 20 minutes of doctor time (and that may be too long in some situations). A person's attention span is approximately 17 minutes, so keep the presentation short, to the point, visual, and motivational.

After you have completed explaining your treatment recommendations and after you have shown him the end results and benefits of the treatment, then go to Step 4.

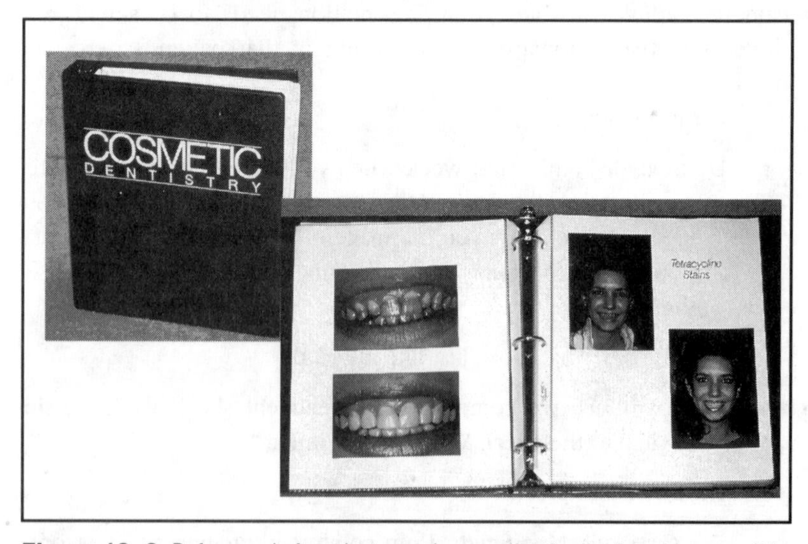

Figure 10–8 Before and after photography excellently displayed is one of the best marketing and educational tools you can develop (courtesy of Dr. Ross Nash).

STEP 4: ASK FOR THE COMMITMENT

Dental teams oftentimes fall short here. They are a little bit uncomfortable asking for a commitment—or closing! Get comfortable with this aspect of case presentation by practicing or role playing various situations as a team. Unless you ask for a commitment, you will have a lot of people walking out the door not knowing if they are going ahead with treatment or not. So it is appropriate to ask for a commitment from the patient.

I have so many doctors tell me that they will be mowing the lawn or watching TV, or sleeping and all of a sudden they wonder, "What ever happened to good ol' Mrs. Jones? I wonder if she scheduled an appointment. I wonder if she is going ahead with treatment. I have to remember to ask Susie about her tomorrow." Then they stew about the patient. Folks, this is a stress point deluxe!

Get comfortable asking for a commitment. You must become comfortable asking for this commitment for two reasons: (1) to find out if

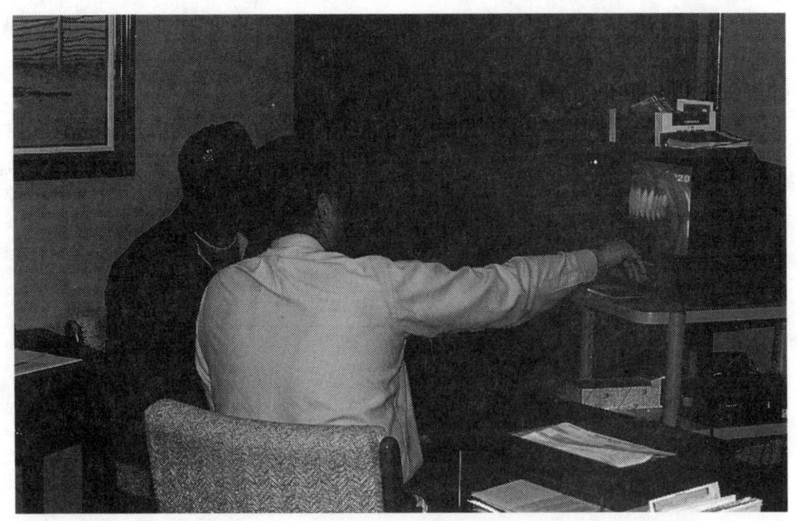

Figure 10–9 During consultation appointments an intra-oral camera will show patients their needs as well as the benefits of possible treatment.

the person is going ahead, or if they aren't, and (2) to uncover any barriers to treatment acceptance. Otherwise nonacceptance will result much too often.

For example:

DOCTOR: Mrs. Jones, have I explained the treatment so that you are comfortable with my explanation?

PATIENT: Yes, I'm clear on what needs to be done.

DOCTOR: Then do you have any further questions?

PATIENT: Well, no, I guess not.

DOCTOR: Then is there any reason why we shouldn't go ahead and schedule an appointment to begin your treatment?

PATIENT: I see no problems. Let's go ahead.

DOCTOR: Okay. I am going to have Pam, my treatment coordinator, discuss the financial responsibilities, and then she will schedule your first appointment. I will look forward to working with you.

By asking the question, "Is there any reason why we shouldn't go ahead and schedule an appointment to begin your treatment?" you are trying to uncover barriers to treatment acceptance. Unless you find out what the barriers to treatment acceptance are, you don't have a chance to clear the way.

Find out what is getting in the way of a person saying yes to treatment. Then you have an opportunity to do something about it. You may or may not be able to clear the way, but at least you have a chance to defuse the barrier. The only way to define the barriers to acceptance is to ask questions.

Three key principles to know about objections:

1. The way to identify an objection is to ask questions.
2. People will not present an objection if they are not interested in a proposal.
3. When people present objections, they are asking for further information.

We'll learn more about handling objections in Chapter 11.

Once the closing question has been asked and there are no more clinical questions, the doctor excuses him or herself and the treatment coordinator or business manager takes over.

The business manager makes financial arrangements (Step 5) and then schedules the appointment (Step 6). There is no exception to this order.

Financial arrangements always precede the scheduling of an appointment.

Special note: In our own practice, our treatment coordinator joins the doctor while he is planning and presenting the case. She is then able to give necessary backup support for the doctor and is able to answer questions that the patient may ask her but doesn't feel comfortable asking the doctor. She is clear about the financial responsibilities and is prepared to discuss appropriate financial options with the patient prior to the scheduling of the first appointment. Careful planning by all team members prior to the consultation moves us toward greater case acceptance.

In Summary

There are the six steps of case presentation/treatment acceptance:

1. Build the relationship
2. Establish the need
3. Instill the desire
4. Ask for a commitment
5. Make financial arrangements
6. Schedule an appointment

Follow this proven six-step process, and you will find that your acceptance rate will increase. The key to this is building the relationship, helping the person feel involved in the decision-making process.

"Tell me and I will forget. Show me and I might remember. Involve me and I will understand."

CHINESE PROVERB

Involve a person in case presentation by asking questions and finding out what he/she wants or needs. Respond to those wants and needs. Use visual aids to more successfully educate. Ask for the commitment. Ask questions to find out what the barriers to acceptance might be. The patient then has a chance to tell you barriers that are getting in the way, and you have a chance to clear them.

This method of case presentation is a two-way communication between you, the providers, and the patient. Both parties benefit: the patient receives wanted and needed dental treatment, and the dental team gets the opportunity to perform great dentistry. Everyone wins! That's dental teamwork!

CHAPTER 11

OVERCOMING OBJECTIONS

"I am not judged by the number of times I fail,
but by the number of times I succeed, and
the number of times I succeed is in direct proportion to the
number of times I can *fail and keep trying."*

TOM HOPKINS

In Chapter 10 (Making an Effective Case Presentation), I encouraged you to ask for a commitment to find out if the patient is ready to move ahead with treatment or if there are any barriers or objections that need to be addressed. Let me reinforce the truism that unless an objection is identified and addressed, the likelihood of a person going ahead with treatment is slim. In addition, if a person goes ahead with treatment, but there is an objection that has not been solved, problems may surface somewhere down the line.

The question is, "How do you define the objections and then how do you deal effectively with them?"

Remember from Chapter 10 the three key principles to know about objections:

1. The way to identify an objection is to ask questions.
2. People will not present objections if they are not interested in your proposal.
3. When people present objections, they are asking for further information.

Don't be afraid of objections. Look forward to them. Look at an objection as a gift. It means that the person is interested. It means that they want you to tell them more. It means that they are asking you to help them make a decision. What a gift! Objections are great!

HANDLING OBJECTIONS

Tom Hopkins of Scottsdale, Arizona is considered one of the best sales trainers in the world. As a student of Mr. Hopkins, I have learned the valuable skills of identifying and handling objections. I cannot think of anything that has been more beneficial to me in my professional and personal life.

Objections, barriers, or problems are going to happen to everyone, probably every day. Success in any relationship and in any business will depend on your ability to effectively deal with these realities, these problems and objections.

When an objection is posed by a patient, take the following steps:

STEP 1: HEAR OUT THE OBJECTION

Don't interrupt. Encourage people to express themselves. Objections often diminish when a person is allowed to talk about it. In addition, this gives you another chance to listen, to show concern, to empathize (not sympathize), and to let the person sense your understanding. Thus you validate your patient!

Let people tell you *everything*. Hear them out. Let them tell you as much as possible. Information is power. The more they tell you, the better chance you will have to solve the problem. Don't become defensive. This will only cause the other person to get defensive, and you will go nowhere. If someone has a problem, don't take it personally. More than likely he/she is upset with the situation, not with you.

Ask questions. Use "listening" body language. Passively listen to encourage them to go on.

STEP 2: ACTIVELY LISTEN

Rephrase and reflect back to the person what you think you have heard them say. This gives you a chance to:

- Clarify
- Reinforce the patient
- Move forward

Pay attention to the feelings being expressed and the behavior that is causing the person to feel that way. This gives you a chance to clarify. "Are you hearing them accurately?" In addition, if the person is angry or upset, actively listening to them will prove to be a calming agent.

STEP 3: REINFORCE THE IMPORTANCE OF THE OBJECTION

There's *no* benefit to disagreeing with or arguing with patients. When you listen to their concerns, express an understanding of those concerns. Share in the development of possible solutions. If you can accurately determine what the objection is without irritating the person, you will be in a position to generate possible ideas for a solution. That's what you want—a solution to the problem. If an acceptable solution can be developed, you will be more likely to see that patient leave with a scheduled appointment.

For example:

Patient: I don't want to lose my teeth, but I sure don't want to spend this much money if this isn't going to last.

Doctor: Keeping your teeth for a lifetime is important to you, and you want to make sure that the investment you make is going to be one that lasts for as long as possible, is that right?

Patient: Yes.

Doctor: I totally agree with you.

STEP 4: ANSWER THE OBJECTION

Provide further education. Stress the end results and benefits of the treatment you are recommending. Turn the objection into a benefit. Establish value. Use the "feel, felt, found" response:

"Mr. Patient, I understand how you feel. Many patients have felt the same concern about making an investment in comprehensive dental care, until they found out that an investment in quality, comprehensive care *now* will:

- Provide better health
- Last longer
- Look better
- Save money in the long run."

STEP 5: OFFER A SOLUTION TO THE PROBLEM

Give patients answers to their questions, or give alternative-of-choice answers to their questions. An alternative-of-choice answer means that you will give the person two options, both of which you like. More than likely—if you have listened carefully—you will be able to offer solutions that will resolve the issue.

PATIENT: I don't know about this. I'm not sure I can afford this. Is there any way I can pay this out for a while?

DOCTOR: We do have several financial options available, and I'm sure one will work well for you. We have options that will let you stretch out your payments over a long period of time. Would you prefer to arrange for three equal payments or do you want to spread the payments out over a longer period of time?

or

DOCTOR: This type of comprehensive care, provided now, would answer your concern about making a stable, long-term investment, wouldn't it?

Check to see if the solution will work for the patient. If the solution answers the question and solves the problem, then you are well on your way.

DOCTOR: If we are able to make the financing of the dentistry comfortable for you, is this the type of treatment you would like to receive?

STEP 6: MOVE FORWARD

Once you have answered the objection, change the flow or focus of the conversation. Move to another area of interest that will move the conversation in a positive direction.

DOCTOR: I was wondering, Mr. Patient, do you have any particular scheduling concerns that we need to be aware of?

STEP 7: CLOSE

Once you have dealt with the objections, ask for a commitment: Close. Closing an agreement means asking!

DOCTOR: Mrs. Jones, is there anything else that would keep us from going ahead with your treatment, or are you ready to schedule that first appointment?

Stay in control:

1. Hear out the objection.
2. Actively listen.
3. Reinforce the importance of the objection.
4. Answer the objection.
5. Offer a solution to the problem.
6. Change the direction of the conversation, move forward.
7. Close.

Remember that you stay in control of a conversation by asking questions—carefully engineered questions that lead the conversation in the direction you want to go. When a person poses an objection, don't freeze up and feel that you have hit a dead end. Not so! As you learn to skillfully handle objections you will find that these objections are progressive steps taken to move ahead.

FOUR MAJOR DENTAL OBJECTIONS: HOW TO IDENTIFY THEM AND OVERCOME THEM

The noncompliant patient! Oh, what energy one must exude to deal with these folks! Do any of the following situations produce stress in your life?

1. Knowing what a person needs, but not being able to lead them to a decision to go ahead
2. Knowing the dentistry is there, but not being able to turn treatment plans into treatment realities
3. Having a patient who does not follow through with preventive home care measures or with necessary co-therapy
4. Dealing with patients who break appointments or who don't show up

Can competency in communication skills help in these situations? Can communication skills be an asset to acquiring case acceptance? or treatment completion? or cooperation with home care? or co-therapy? Can your ability to communicate effectively have an effect on your bottom line? Can communication expertise help control the stress caused by noncompliance? *Yes!*

Dr. Jameson, having practiced dentistry for 20 years, shares the following insight:

> *Not having patients accept treatment recommendations or not having them comply with treatment modalities was a big frustration to me in my earlier practice years. I knew* how *to do the dentistry, but I couldn't figure out why people didn't want it as much as I wanted to do it. Patients weren't as sure about how much they needed treatment as I was!*
>
> *I didn't feel good about myself—or my dentistry. I began to lose my enthusiasm. Then I discovered that I was missing the boat. I didn't know enough communication skills to get the message across. I was simply—and ineffectively—telling people what they needed. I wasn't determining motivators, and I wasn't thinking about problems and solutions. I could talk, but I wasn't communicating.*
>
> *When I made the effort to study and use communication skills, everything changed for the better. Now, people are following through with comprehensive care. They are happy—and so am I!*

Let's look at some stressful situations and apply the communication skills we have learned so far to work at gaining better control of the situations.

You, as the dentist, know what a person needs in his/her mouth to accomplish excellent oral health and/or an attractive smile. Frustration comes when you see what a patient needs, and you can't encourage them to go ahead. Excellent presentations are critical for treatment acceptance.

We have just studied the six effective steps for achieving excellent results from your presentations:

1. Build the relationship. Develop a comfortable level of trust and confidence.
2. Establish the need. Uncover both the emotional need and the diagnostic need.
3. Instill the desire. Educate and motivate!
4. Ask for a commitment. Close.
5. Make financial arrangements.
6. Schedule the appointment.

Many times, the treatment recommendations are explained; the doctor or treatment coordinator asks if there are any questions; the patient says no; and that is that. There is no asking for the commitment. Then doctors wonder, "What happened to Mrs. Jones? Did she schedule an appointment? She didn't? I wonder why?"

Find out why before Mrs. Jones leaves the consultation appointment. If she is not going to comply with recommendations, uncover the objections or barriers before she leaves the office. Then—and only then—will you have the opportunity to resolve or defuse barriers to treatment acceptance.

How can you do that? How can you find out "why"?

Once treatment recommendations have been presented (in layman's terms), it is appropriate to ask if there are any questions. But control those questions. First find if there is any confusion about the treatment itself. Then begin your search for barriers.

For example:

DOCTOR: Mrs. Jones, have I explained my treatment recommendations so that you are comfortable with my explanation?

or

DOCTOR: Mrs. Jones, do you have any questions about the treatment itself?

These questions focus the patient on the treatment and on questions about that treatment. You will want to deal with technical questions before entering into a discussion of potential obstacles such as finances.

These questions also place the responsibility of the explanation right where it belongs, and that is on the teacher: the doctor or the treatment coordinator. *You* are responsible for the explanation—the patient is not responsible for the understanding. *You* are the teacher.

Once any confusion or questions are cleared, the next step is to investigate possible objections.

You will be asking for a commitment to proceed with treatment, and you will be trying to isolate barriers at the same time. This is a tremendous opportunity. If you never know what the barrier to treatment acceptance is, you will never have a chance to overcome that barrier.

Overcoming Objections

Remember the four main reasons that people do not comply with your treatment or prevention recommendations? They are:

1. No perceived need, lack of dental education
2. Fear of cost
3. Fear of the dentistry itself
4. Time

If you can constructively deal with or overcome these four barriers, your number of noncompliant patients will decrease.

Tom Hopkins identifies the four Ps of professionalism as:

1. Preplan
2. Practice
3. Perfect
4. Perform

Mr. Hopkins suggests that if these four steps are followed, any skill can be mastered. How can you use these to overcome the four major objections? As a team do the following:

1. Preplan
    - Discuss each of these four objections, citing specific examples of each.
    - Plan how you will try to overcome those objections.
    - Write verbal scripts.
2. Practice
    - Practice those scripts with each other. Role play. Get comfortable.
    - Practice the skills as patients flow through the office and as objections are presented.
3. Perfect
    - Perfect your skills. Evaluate various situations.
    - Write down and/or discuss what went well and what didn't go well.
    - If something is working well for you, stress this point to the other team members. Do more to build on these positives.
    - If something is not working well for you, then change it! Redo, rewrite, replan your course of action. But don't throw in the towel if your first plan doesn't work! Persevere.
4. Perform
    - Continue to make your presentations even during that tough period of change.
    - Don't fall back into your old ways. Continue to move forward and to get better.

LISTEN ACCURATELY

Remember active listening? When a person poses an objection, reflect back to the person what you think you hear to make sure that you heard correctly. This type of listening is critical if the objection is to be identified and overcome.

By actively listening or responding reflectively you accomplish the following:

1. Establish empathy
2. Express a caring attitude and warmth
3. Show respect
4. Defuse fear or anger
5. Move the conversation forward

Ultimately, you stay in control of a conversation by asking questions and then accurately listening.

Here are suggested communication/verbal skills for dealing with these four main reasons for noncompliance or the four major objections in dentistry. Notice the use of questions and active listening.

Situation 1: No Perceived Need

A patient is informed of the presence of decay following a continuous care appointment.

MS. PATIENT: I'm sick of hearing that I have decay every time I come in for a cleaning.

HYGIENIST: You're bothered by the fact that we find decay at most of your visits?

MS. PATIENT: Yes! I'd love to come to have my teeth cleaned and be able to say, "Look, Mom, no cavities!"

HYGIENIST: I'm confused. You say you'd like those "no cavity" appointments, but the status of your mouth shows that your home care is not regular. Are you saying that in order to become healthier and to prevent the decay, you would become more committed to a co-therapy regime?

Softly, but effectively, the hygienist identifies a discrepancy in the patient's values. This discrepancy and the results of such a conflict are revealed. Then the patient is given the opportunity to decide if she wants to make the necessary changes. If she makes an affirmative decision, she will be more likely to stick to it than if the hygienist decided for her. You can't push a person into making a decision, but you can lead them to that decision.

Situation 2: Fear of Cost

PATIENT: Well, Doctor, I'd like to go ahead with this treatment. I know I need it. But it just costs too much! I can't afford it right now. I'll just have to wait.

Have you heard this before? Does the response ever come at the completion of your excellent presentation of recommendations? Do you get discouraged? Do you wonder what you can do to deal with the objection, the barrier, the fear of cost?

Let's look at a step-by-step way to handle the objection of cost. It's a biggy!

1. Validate your services and your quality to yourselves.

   Do you feel that the value of your services exceeds the fee you are asking for it? Before anything else happens, you must convince yourself of your own worth! The confidence that you and every member of the team have in the quality of and the benefit of the treatment that you are offering are of central importance to overcoming the fear of cost.

   Key Point: Your entire team *must* believe in the services you are providing.

   Everyone must have a strong commitment to their work and to the patients they serve. You, as dental professionals, add value to the lives of those people. Be constantly clear about the mission and purpose of your practice (see Chapter 2). If you do not already have a written mission statement, I encourage you to have a team meeting where you discuss the following:

   1. Who you are
   2. What you are about
   3. What you intend to accomplish in your practice

   The statement of mission needs to be empowered by emotional words that are motivational to you. Everyone needs to agree that the statement says it all. The statement should be short and to the point, but powerful in its message. Have the message printed, matted, and framed. Hang it where everyone can see it on a regular basis.

As you are writing your goals for your practice, you need to ask yourself this question, "If we accomplish this goal, will it bring us closer to satisfying our ultimate mission?" If the answer is yes, then you probably have a good goal. Go for it!

Make sure that the treatment your patients are receiving is an equitable exchange for the fee. Thus Value = Value.

Exercise: List the services you provide for your patients, from the initial contact through the entire treatment.

1. What makes your services special?
2. What added-value touches do you provide that make your practice unique?
3. What do you do that goes beyond the expected?

Do this exercise as a team. This will help you to see in your own hearts and minds the things that you are presently doing in which you go above and beyond. This will make you see the value of your services and will help you deal with the issue of cost within yourself. Remember: This is the first step to overcoming the fear of cost in your patients—validating the fee for the services in the minds and hearts of all team members.

2. Validate yourself personally to your patients.

Key Point: You must establish a relationship of trust and confidence with a patient before treatment acceptance will result.

Patients must know that they can count on you and can receive the same type of treatment and care each and every time they are in your practice. Consistency of care is critical to establishing the trust that is foundational to treatment acceptance. Your ongoing internal marketing program or PR program should have this as its foundation.

In planning your marketing/educational program, ask this question: Does this marketing tool make a statement (consciously or subconsciously) about who we are, what we do, what our purpose is about? If the answer is yes, then the marketing tool is probably going to serve your purpose well. If the answer is no, then you may need to rethink the project.

3. Validate your services.

In your efforts to validate your services to existing and potential clients, let patients know the following three things:

1. Description/Explanation: What it is you are recommending. (For example, What is a bridge?)
2. Advantages: What the benefits are of this procedure. (How will this help the patient?)
3. Proof: Can you produce the results?
   - Use testimonial letters from enthusiastic patients.
   - Use before and after photographs of your patients to illustrate a particular service you provide. (Be sure to obtain written permission from your patients for use of photography.)
   - Provide civic presentations throughout your community using before and after slides of treatment you have provided. Of course, this presentation would be nonself-serving. It would be an educational program letting people in your area know what is happening in dentistry, what kind of services are available to them today. However, tremendous credibility can result.
   - Intra-Oral Cameras—Using an intra-oral camera to show a person what they have in their mouth presently validates your recommendations more than anything else. When people can see the evidence for themselves, they will see the needs immediately.
   - In addition, before and after images stored on disk can be recaptured to provide the necessary proof that you can accomplish the described results.

4. Make sure that every aspect of your practice epitomizes the professional image you wish to project.

Key Point: You want to have the exchange of value be one of equability, but one that is perceived to tilt in the favor of the patient.

Be committed to comprehensive quality—quality in every fiber of your practice. Make sure that everything in your practice consistently sends a message of excellence to your patients—from the telephone

Figure 11–1 Participation in community or civic presentations spreads the word about good oral health. Exposure to the community results.

conversations, to the facility, to the written communication, to the actual treatment, to the management situations, to each and every team member. Evaluate. Be honest with yourselves. What says *quality?* What doesn't?

Drive up to your own office. Put on your "patient eyes." What impression do you get? Does every step of the patient visit invite trust and confidence? Does everything speak of quality? Are the patients going to get the impression that you will take good care of them by the kind of care you give your facility? Is the team focused on one thing: the care of the patient? Does that come across loud and clear?

Handling the Cost Objection

By doing the above four steps you begin to neutralize or overcome the fear of cost, or the fear of buying.

An objection—including the objection of cost—is actually a step forward in completing an agreement. If your patients do *not* pose any objections or raise any questions, they're probably not interested.

Remember: An objection is actually an opportunity for you. It defines a specific area of concern. In fact, you *need* to ask questions to isolate or to identify what objections, if any, might get in the way of a person going ahead with treatment.

Key Point: If you know something is going to be brought up as an objection, *you* bring it up. This gives you an opportunity to turn a potential negative into a positive in advance.

DOCTOR: Mr. Patient, before I give you the results of my analysis, and before I explain the treatment that I am going to recommend for you, first let me tell you that if you have any concerns about the financing of your treatment, we do have convenient, long-term financing right here in our office. I tell you this so that for now we can both concentrate on the treatment I am recommending. We will discuss financial options in full. We want to make sure that you are clear and comfortable with this important part of your treatment. Okay?

After you have made your presentation of recommendations and answered all questions about the treatment, if the patient expresses a concern about the money, then ask this question:

DOCTOR: And so, Mr. Patient, the financing of the dentistry is a concern for you, is that right?

PATIENT: Yes.

DOCTOR: Other than this, is there any reason why we shouldn't go ahead now?

PATIENT: No, I want to do this. I just need a way to pay for it.

or

DOCTOR: Mr. Patient, if I understand you correctly, this is the type of dentistry you would like to receive, is this right?

PATIENT: Yes.

DOCTOR: Then if we are able to make the financing comfortable for you, is there any other reason why we shouldn't go ahead and schedule an appointment to begin your treatment?

PATIENT: No.

Objections diminish when a person is allowed and encouraged to talk about them. And so

1. Restate the patient's wants and needs.
2. Actively listen to their concerns: rephrase and feed back their objections.
3. Validate the person.
4. Turn the patient's objections around by asking a question to establish value.
5. Encourage the patient to share with you in the development of a solution.

Key Point: "If people are allowed to be a part of the decision-making process, they will be more likely to buy into the decision." MICHAEL DOYLE

You can't push anyone into making a decision, but you can lead them by carefully and caringly asking questions and listening. You can't *talk* people into going ahead, but you can *listen* them into going ahead.

Common Financial Objections

Objection 1. "That costs too much."

BUSINESS ADMINISTRATOR: You feel the fee is too high for the services we are recommending for you? Or is the investment difficult for you at this time?

PATIENT: I'm sure the treatment is worth the fee, but I can't afford this right now. That just costs too much!

BUSINESS ADMINISTRATOR: I know how you feel, Mr. Patient; today most things do. Tell me, Mr. Patient, if we can make the financing comfortable for you with a convenient monthly payment plan, would this make it possible for you to proceed?

PATIENT: Probably.

BUSINESS ADMINISTRATOR: How much per month could you invest?

His answer to this question would let you know if you could go ahead by offering him MasterCard, Visa, Discover, or a health-care financing program.

or

PATIENT: I want those veneers. I hate my smile. But $3,000 is just too much!

TREATMENT COORDINATOR: How much too much is that, Mr. Patient?

PATIENT: About $1,500 too much. I saved $1,500 for this, but I had no idea it would be this much!

TREATMENT COORDINATOR: So the solution we're looking for is a way to finance the $1,500 beyond your savings program, is that right?

PATIENT: Yes.

Now you know that the $3,000 isn't the problem—it's the $1,500 of the treatment that needs attention and assistance.

Objection 2. "I'll have to think about this!"

DOCTOR: Thank you, Mr. Patient. I'm sure that you wouldn't take the time to think about the treatment if you weren't interested. In order for me to be clear as to what you need to think about, let me ask you if it's the treatment itself you need to think about, or is this the type of dental care you'd like to receive?

MR. PATIENT: I want the treatment. I know this is what I need! I'm sick of the mess in my mouth.

DOCTOR: Then, are you unsure about us providing your treatment? Are you unsure about me?

MR. PATIENT: Oh, no. I wouldn't have come here if I didn't trust you guys!

DOCTOR: Well, may I ask, is it the investment? Are you concerned about the money?

MR. PATIENT: Yes! I had no idea it would cost this much to get my teeth fixed! I don't know if I can afford this right now.

DOCTOR: If we are able to make the financing of the dentistry comfortable for you, would that make it possible for you to go ahead with the recommended treatment?

MR. PATIENT: Well, yes. If I can pay it out.

DOCTOR: If we are able to arrange comfortable financing, is there any other reason why we should not proceed with your treatment?

MR. PATIENT: No, not really.

DOCTOR: Mr. Patient, I'm going to ask Judy, my business manager, to spend some time with you discussing our financing options. She is excellent, and I know that if anything can be arranged, she will do just that. The last thing we want is for the fee for the service to get in the way of you receiving the treatment you need.

This is a closing sequence in which the doctor responded to the patient's questions or statements with another question or with a clarifying response. The doctor didn't jump to any conclusions. He/she didn't offer solutions, make judgments, or give advice. The doctor did not become offended by or angered by the patient's concern about the fee.

This closing sequence does the following:

- Gives the patient a chance to uncover the barriers
- Lets the doctor clarify how many and what kind of obstacles are present
- Asks a closing question that leads a patient to make a decision

The problem is not totally solved! The doctor has "preclosed." The financial coordinator must now step in to try and make comfortable financial arrangements and then to ask for the final commitment.

Objection 3. "If my insurance doesn't cover this, I can't get it done."

BUSINESS ADMINISTRATOR: Mrs. Patient, it is great that you have dental insurance. Our patients have found that dental insurance is a wonderful supplement to their health care. However, insurance is not a pay-all. It is a supplement. Dr. Jameson has recommended treatment that is necessary for the restoration of your mouth. We will do everything we can to help you maximize your insurance benefits. Based on the verification we have obtained from your insurance company and on the experience

> we have with them, we can estimate very closely what we expect them to pay. You will be responsible for the balance. However, we want you to know that we have excellent payment options available in our practice and can assist you with that balance.
>
> Mrs. Patient, how much per month would you be willing and able to invest?

I have spent time here dealing with this issue of fear of cost. Why? Because it is such a difficult situation for the dental professional to handle and one that seems to be ever present. These skills are tried and true. You have to decide—Is the result of overcoming the objections, allowing more people to say yes to your treatment recommendations, increasing production, and letting more people receive your excellent care worth the effort? I hope so. Preplan, Practice, Perfect, Perform!

Exercise:

1. List the main financial barriers or objections your patients give to you.
2. Using the skills from this chapter, formulate scripts that will help you to deal with and overcome those objections.
3. Role play using these scripts.
4. After the role playing, answer these questions.
    - Did I listen carefully?
    - Did I reflect back what I thought were the patient's main concerns?
    - Did I validate the patient?
    - Did I answer each objection with a "values" question?
    - Did I go through the steps of dealing with an objection?
    - When I overcame the objections, did I close?

Situation 3: Fear of the Dentistry Itself

For example: A patient is presented with ideal dentistry, but her dental fear is getting in the way of her going ahead with the treatment of choice.

MS. PATIENT: Root canal! No way! I've heard about those things! I just want you to pull that tooth out.

DOCTOR: Sounds like you're apprehensive about the treatment I've recommended for you.

MS. PATIENT: That's putting it mildly! I don't want to lose that tooth, but I don't think I could handle a root canal. (Use the "feel, felt, found" method here and use great visual aids to educate.)

Doctor: I understand how you *feel,* Ms. Patient. Many of our patients have *felt* the same way, until they *found* out that we are committed to comfortable dentistry here, and that we will make sure your treatment is as short and as comfortable as possible.

Since you want to save the tooth and since we *can* do this in a timely and comfortable manner, is there any other reason why we should not go ahead with your treatment?

The doctor didn't impose a value judgment on the patient, such as, "That's silly! I can't believe you want to have this tooth extracted." or "There's no reason to be nervous/scared/worried about a root canal!" It does no good to *tell* people not to be afraid. If they are, they are! Acknowledge the fear. Validate their emotions. Then make efforts to educate, resolve the fear, and build confidence. You'll find that your warmth and empathy will take you a great deal further. If you can deal positively with the fear of the dentistry, you'll have many more compliant patients and fewer broken appointments and no-shows.

In his book *Helping and Human Relations: A Primer for Lay and Professional Helpers,* R. R. Carhuff points out that professionals who communicate with a high degree of warmth, caring, and empathy are reported by patients to be more effective in social, personal, and vocational functioning. This type of communication makes it easier for people to openly discuss problems, express feelings related to a situation, and to make decisions about treatment. In other words, compliance and cooperation are advanced.

Situation 4: Time

For example: It's not convenient for a patient to schedule appointments, keep appointments, or to commit to full, comprehensive treatment.

APPOINTMENT COORDINATOR: Mr. Patient, we reserve specific times of the day for this type of intricate and detailed procedure. In order for the doctor and the clinical team to give the necessary attention to you, we need to reserve special time. Tell me, are mornings or afternoons best for you?

MR. PATIENT: Mornings are best, but I don't know if I can be here next week. Can I just call you when I can come in?

APPOINTMENT COORDINATOR: Next week is a problem for you?

MR. PATIENT: Yes. I have meetings all week and I just don't think you could count on me.

APPOINTMENT COORDINATOR: Well, I appreciate your honesty. I don't want to reserve this time for Dr. Jameson without being sure you will be here! Does the week after look better for you?

MR. PATIENT: I just don't know. I'll just call you.

APPOINTMENT COORDINATOR: Mr. Patient, let me make a note to myself that you do wish to proceed with this treatment, but that time is a concern for you. With your permission, I will call you in a couple of weeks to schedule the first appointment. Would that be okay with you?

MR. PATIENT: Sure.

The appointment coordinator maintained control of the conversation and of her appointment book by asking the patient's permission to call him back. When you let patients say that they are going to call you, you have just turned the controls over to them.

In this type of situation where the patient is difficult to schedule or cannot be counted on to keep appointments, it is vital that you do the following:

1. Maintain control of the situation.

2. Set the tone for the appointment by stressing

    - The importance of the appointment
    - That you reserve special time for them

- That because they are special you want to give them your best care
- Then offer appointment times

3. Offer alternatives of choice. Offer two choices, no more. Either one of the choices is to your liking. (For example: mornings or afternoons, Tuesday at 9:00 or Thursday at 11:00.)
4. Place the responsibility of the appointment on the shoulders of the patient.
5. Ask permission to call a patient rather than letting them call you. Oftentimes when patients tell you that they will call you, they never do. Stay in control. Ask permission to call them. Then make a note in your tickler file to *do so!*

The appointment coordinator stayed in control of this conversation and of the appointment book. When you allow the patients to control your days, their appointments, and, ultimately, your practice, chaos can result. Chaos breeds stress.

Create a Win/Win/Win/Practice

You do your patients a tremendous service when you control your practice, because everyone then becomes a winner. Your patients win because you can give them your full attention. Your team members win, because they aren't running around in chaos and can turn their attention to patient care and to maximizing their own talents. You win because

1. You are able to provide optimum dental care more often.
2. You can function in a stress-controlled environment.
3. You can take the necessary time to build and maintain excellent relationships.
4. You can enjoy the practice of dentistry and love what you do for a living!

AND FINALLY . . .

Do not fear an objection—even the objection of money. Rather, look at this as an opportunity.

Know that if you do your best and if people do not go ahead, they are rejecting the treatment proposal. They are *not* rejecting you.

Combine a strong belief in your team and the services you provide with the skills to get that message across. Then you can effectively overcome objections.

CHAPTER 12

NURTURE PERSONAL REFERRALS: YOUR BEST SOURCE OF NEW PATIENTS

"Many people refuse to delegate to other people because they feel it takes too much time and effort and they could do the job better themselves. But effectively delegating to others is perhaps the single most powerful high-leverage activity there is."

STEPHEN R. COVEY

Without question, your number one source of new patients has been—and always will be—personal referrals. In fact, if you are in general practice, approximately 70% of your new patients will come to you as a result of personal referrals. If this is the case, then everyone on the team must be given and must receive the responsibility and the commission to nurture those referrals. By doing so, you will be delegating the responsibility for building your practice to all members of the team and to the hundreds of patients who are already a part of your organization. As Mr. Covey says, that may be the "single most powerful high-leverage activity there is."

Throughout this book, we have studied how to communicate excellently so that your patients will recognize your professionalism and your commitment to quality. We have learned how to deal with difficult people and difficult situations so that you can turn those challenging situations into great relationships. We have studied the intricate skills of case presentation so that people will accept the treatment that you are recommending.

Now you will want to continue the flow by turning those happy patients into referral sources: people who willingly recommend you to their friends and family members.

IDENTIFY AND NURTURE REFERRAL SOURCES

ASK FOR REFERRALS

Everyone on the team needs to become comfortable with asking for referrals. When a patient expresses gratitude, has completed treatment, has received a beautiful new smile, etc., ask that person for referrals.

For example:

MRS. JONES: I just love my new smile. I feel so great!

DOCTOR OR TEAM MEMBER: Thank you, Mrs. Jones. We appreciate your kind words. Mrs. Jones, we build our practice on excellent patients like you, and we love creating those beautiful smiles. Do you have friends or family members who might need a dental home?

MRS. JONES: Oh, yes. Lots of them.

DOCTOR OR TEAM MEMBER: May we give you a few of our cards (or office brochures) and ask that you tell your friends and family members about our practice? When they come to us, we will take excellent care of them, just as we have tried to do with you.

People will be happy to refer their friends to you. Ask! Believe in what you are doing and ask your solid clients or patients to give their support to you.

ACKNOWLEDGE REFERRAL SOURCES

Business experts tell us that one of the most effective ways to solidify or expand a relationship with a client is to write notes. As simple as this may seem, it works. Acknowledging and expressing your gratitude goes a long way.

Either produce your own note or purchase notes from a supplier. Keep these notes in the business office with a specific person assigned the responsibility for completing and tracking these notes.

Sign and address these cards every day, and the task becomes manageable. Leave them to do all at once, and all of a sudden it becomes an overwhelming task—sometimes left undone! Keep this effort simple! If the task doesn't become overwhelming, it will be done on a regular, consistent basis.

In our quarterly newsletter, we also acknowledge and thank the

DR. JOHN H. JAMESON AND HIS STAFF

Thank you for referring

____

to our office for dental care.
We are grateful for the confidence
you have shown in us by referring
your friends and family to our office.
We appreciate you and the people you refer.

101 JAMESON DRIVE
WYNNEWOOD, OKLAHOMA
(405) 665-2041

Figure 12–1 Always acknowledge your referral sources. This simple, but effective courtesy is appreciated by those valuable patients who make referrals.

patients within the practice who have referred a new patient to us. Recognition is a major motivator, according to Harvard University. And so, we recognize the referral sources so that we can encourage them to do this again! We are positively reinforcing this excellent behavior of referring. Positive reinforcement is the best way to solidify behavior. "That which is rewarded is repeated," according to Dr. Michael LeBoeuf, head of management for the University of New Orleans.

GIVE GIFTS AS AN EXPRESSION OF THANKS TO YOUR REFERRAL SOURCES

Consider providing an added-value benefit to your new patients. Involve the referrer in this added value. Added-value marketing is a proven incentive to a new client.

Produce a gift certificate that you can place in the hands of your regular patients. The patient of record will be the giver of the gift. Ask your patients to give this gift certificate to their friends and family members. You can make the gift whatever you wish. Some offices give a complimentary examination and consultation. Other offices give a variety of gifts or incentives—perhaps a percentage fee reduction following a certain number of referrals.

Complimentary Oral Health Supplies

In our office, we distribute certificates called "A Gift of Health." Everyone on the team gives these cards to existing patients and asks those "satisfied clients" to refer friends and family members to the practice.

The gift certificate is for a six-month supply (12) toothbrushes to be received upon completion of a comprehensive, initial exam. On the certificate is a brief synopsis of the recent studies by Dr. Richard Gloss of the University of Oklahoma Dental School about the need to change toothbrushes every two weeks. With this certificate and the corresponding gift, we are providing "added value" marketing, and we are educating people about excellent health (see Figs. 12–2 and 12–3).

When a person refers a certain number of people or when a certain amount of dentistry is provided as a result of those referrals, we give that referral source a complimentary oral hygiene device, such as an oral irrigation system or an electric toothbrushing device.

A GIFT OF HEALTH

JOHN H. JAMESON and HIS STAFF

want to offer you a complimentary
six month supply (12) of toothbrushes.

"A recent study has concluded that it is necessary to change toothbrushes every two weeks due to the invasion of bacteria into the bristles and the handle of the brush. Complications of use of a toothbrush longer than two weeks include: mouth sores, fever blisters, colds, gingivitis, and various illnesses.

This certificate is redeemable
at the time of
your comprehensive examination.

Figure 12–2 The gift of health card is an example of added-value marketing (front view).

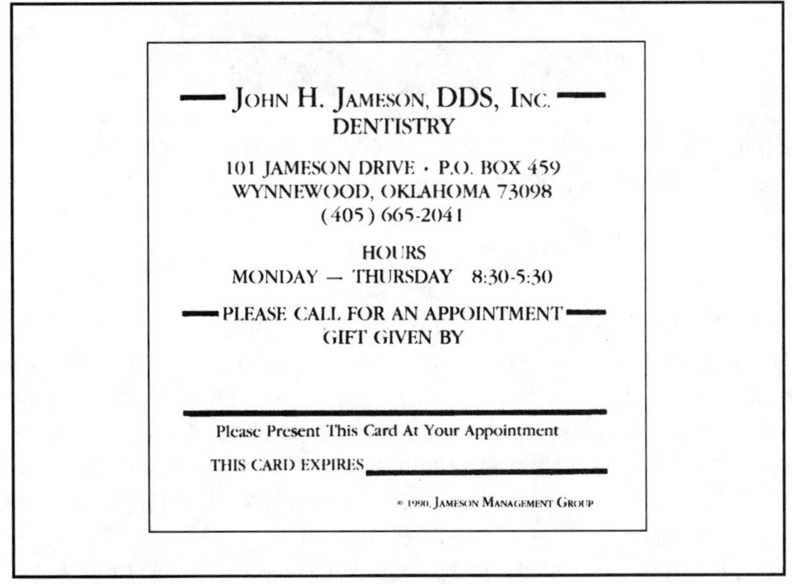

JOHN H. JAMESON, DDS, INC.
DENTISTRY

101 JAMESON DRIVE • P.O. BOX 459
WYNNEWOOD, OKLAHOMA 73098
(405) 665-2041

HOURS
MONDAY — THURSDAY 8:30-5:30

PLEASE CALL FOR AN APPOINTMENT
GIFT GIVEN BY

Please Present This Card At Your Appointment

THIS CARD EXPIRES____________

© 1990, JAMESON MANAGEMENT GROUP

Figure 12–3 The gift of health card (back view).

Figure 12–4 Acknowledge birthdays and other special occasions or say thanks by sending balloons or flowers to a patient at their place of employment.

Coffee Mugs

When a patient refers three people to our practice, we send a coffee mug with either thank-you balloons attached or filled with a flower arrangement. We have had the mugs imprinted with our logo on one side and our name, address, and phone number on the other. We keep these at a local floral shop. We call the shop to place the order. Then we send this thank you to the referring person's place of employment (see Fig. 12–4).

It is our hope that the referrer will carry that mug around the office, happily sipping on coffee and inadvertently placing our name in front of lots of people. Another marketing principle is at work here:

You need to let people know who you are, where you are, and what you do. And, people need to see your name over and over and over.

Repetition is an essential ingredient of good marketing: marketing that gets results!

Complimentary Theater Tickets or Gift Certificates of Dinner for Two to Solid Referrers

Many theaters and restaurants will give you special rates if you make an arrangement to refer on a regular basis to them. Schedule a time to discuss the possibilities with the owner or manager of the facility. Get all necessary paperwork or supplies.

When a person earns this wonderful gift, send a special note on your letterhead stationery that includes the gift certificate. Keep track of what you are sending and how much you are sending. It may benefit you to send these reports to the manager from time to time so that he/she is aware of the health of your business to them. They will, then, become a referral source for you. In fact, you may begin to see patients coming to you that are employed in the restaurant or theater.

Compromise Your Fees

If a person continues to refer people to you who become actively involved with treatment, you may choose to thank them by compromising your fees. You may give them a 5%–7% gift off their next appointment for having sent you a significant amount of new business.

Again, following the lead of excellent business leaders, know where your business is coming from and nurture that. When you scratch their backs, they will scratch yours. People love the recognition and everyone loves a fee break. Track your referral sources and track the amount of dentistry that is produced as a result of their referrals.

NETWORK WITH OTHER PROFESSIONALS

Identify businesses in your area that attract people with similar backgrounds of your patient family. (For example: physicians, pharmacists, bankers, optometrists, retailers, etc.) Schedule an appointment with the leader of the organization and discuss the services that you offer. Ask if you can leave brochures about your services with them, and ask that they refer people to you who might need the services that you are providing.

Spend some quality listening time discovering new information about the services that *they* are offering. Enthusiastically—and sincerely—let them know that you will refer people to them who might need their services. Then, do so, make sure that when you send one of your patients to another professional or business that you send a card with the patient. Ask that they give the card to the particular professional or company so that they will know that you have referred them.

Stay in close contact with these other professionals. Visit with them on a regular basis. Let them know that you want a solid long-term relationship with them. Make personal visits to their practices or places of business and invite them to come to your facility. (These contacts do not have to be made by the doctor. Other members of the team who are interested in and are comfortable with public relations and marketing can be excellent ambassadors for the practice.)

Many of the specialists that I work with have special lunches for their referring practices. The members of the teams like to get acquainted just as much as the doctors. In addition, these lunches will give you a chance to tell the referring practices what you are doing and how your services might benefit their clients.

Be ever aware of the fact that the members of the team can be as influential in referring a patient as the doctor. They will have solid relationships with the patients, also. The patients will listen to and will value their opinion and their statements of referral and confidence.

In addition, specialists will often sponsor an educational seminar for their referring doctors and their teams. This complimentary seminar (which usually includes lunch) will be a powerful way to say thanks and to continue the referrals.

Stay in front of those referral sources. That is one of the most critical of all factors underlying a successful referral system.

Interact with Other Appearance Specialists

Place information about the services of other appearance specialists in your office. Network with these people. Once a person has completed a certain type of treatment, such as full mouth reconstruction or cosmetic dentistry, send them to a salon for a makeover or to a photography studio for a "sitting" or give them a one month membership to a health spa. The money that you will invest will serve as a "thank you" to the patient, but it will also stimulate referrals from the appearance company.

Be sure that they know you are referring. Make the contact and the arrangements yourself, so that you are well-known as a healthy referral source to them. They will be happy to reciprocate.

DESIGN A PLAN THAT WILL WORK, THEN WORK THE PLAN

In your morning meetings, identify those patients who would be great people to ask for referrals—patients who have expressed pleasure in receiving treatment from you. Then, determine who is going to ask for that referral. You wouldn't want to have everyone ask them for a referral, so give someone that responsibility.

Be sure that you have done some role playing and have practiced the verbal skills of asking for a referral. Get comfortable asking. Know that you will never know what you will get unless you ask! All good businesses ask for referrals from their best customers. Learn from the masters. Pattern your own behavior after those who have acquired admirable results.

When you identify patients who have accepted and have been pleased with your treatment, patients who have come to their appointments—on time, patients who have paid their bills (and have been glad to do so), and patients who have been a joy to treat, ask *these* folks for referrals. More than likely they will refer others to you who are of like character.

Together, as a team, develop ways of acknowledging and nurturing your referral sources. If these wonderful people provide more than 70% of your new patients, it benefits you to invest time and money in maximizing this source of practice growth.

CHAPTER 13

STRESS CONTROL: CRITICAL FOR EXCELLENT COMMUNICATION

"You can become productive without being self-destructive."
ROBERT ELIOT, M.D.

In order to be an excellent communicator, you need to be in control of the stresses in your life. Otherwise, you will not have the available attention or the necessary energy to concentrate on the skills. The dental environment can be extremely stressful and that, in itself, can take away from your ability or your willingness to communicate. If you are overwhelmed with stress in your life, it may be easier to ignore or avoid difficult people and difficult situations. If stress becomes a dominant force in your life, you may be so short tempered that you don't have the available patience to communicate when you or the other people in your life need it.

Do you have any stress on the job? If you answered yes, then welcome to the modern-day workplace! Stress is a fact of life in business and in dentistry today. A certain amount of stress is good! It can be motivational. Without a certain amount of stress, success and achievement would be diminished. However, too much—or uncontrolled—stress can be detrimental to the health and well-being of individual members of your organization and to the overall health of the organization itself.

What Is Stress?

In *Dental Practice Management Encyclopedia,* Dr. Carl Caplan says the term *stress* refers to "the physiological and psychological responses of an individual to demands from the environment (termed stressors)." William H. Hendrix, Ph.D., defines stress as "an individual's cognitive interpretation (or appraisal) of internal or external events judged (consciously or unconsciously) to be threatening, harmful, or challenging."

This definition indicates that different stressors affect different people in different ways. A situation that might be perceived as stressful to one person on your team might not be at all stressful to another.

Peter Hanson, M.D., in his book *Stress for Success,* states

> *"Stress is 80% of all illnesses. On the other hand, stress is also the key to excellence. Stress does not actually cause excellence, nor does it actually cause illness or financial losses. In fact, stress is neutral until it lands on a person. What that person has chosen to do about past stresses, and what the person chooses to do in response to present stress, will determine the outcome."*

Dr. Hanson points out that

> *"the answers [to the problems of stress control] are within the grasp of each individual. Each of us, as a manager of his or her own Department of One, has the power to break out of the lemming herd and turn away from the precipice that awaits the incompetent stress handler."*

Uncontrolled stress can affect the mind. It can take its toll on emotional well-being while it drains energy and vitality from you. This can lead to irritability and quick temper. Stress can distort the way you think and feel about yourself. Stress, unbridled, can chip away at your own feeling of self-worth—your self-esteem. Stress can affect your relationship with others, making it difficult to relate in a constructive way with people at work and outside of work. Stress can also drain the energy you need to participate in activities in your community.

According to Dr. Hendrix, there are three main factors that lead to perceived stress among professionals in the workplace today:

1. Job-related factors
2. External, or non-job-related factors
3. Individual or personal characteristics

How do these three main factors affect the practice?

JOB-RELATED FACTORS

The organizational climate of a dental practice is substantially related to stress. The better the organization of the practice, the more controlled the stress levels for all team members. Poor organization—or total lack of organization—is a major source of stress for all involved.

Organization and management can be divided into three specific areas:

1. Business management
2. Personnel management
3. Patient management

Specific organizational or management deficiencies that lead to stress in each of these three areas may be:

1. Business Management
    - Lack of procedures, policies, or systems for dealing with the business aspects of the practice, i.e., scheduling, financial arrangements, insurance management, time management
    - Equipment problems or insufficiently equipped treatment rooms

- The working environment itself: the facility
- Cash flow: profitability or lack of profitability

2. Personnel Management
    - Interpersonal conflict
    - Poor leadership
    - Burnout
    - Staff turnover
    - Ineffective communication
3. Patient Management
    - Managing difficult or fearful patients
    - Dealing with nonacceptance of treatment recommendations
    - Patients who refuse to maintain oral health following treatment
    - Patients who cancel appointments with little or no notice
    - Patients who refuse to pay their bills

In addition to these organizational and management issues, the physical aspects of performing the dentistry also ignite the stress response. The performance of dentistry is extremely tedious. All five senses are constantly stimulated. There is a perceived concept that there is no margin for error, thus imbedding the perfectionist syndrome.

The environment in which the clinical team works is extremely small. The body position that must be assumed is physically taxing and adds stress and strain to nerve and muscle alike if held over a period of time. Additional environmental factors that might add to stress include room temperatures, loud noises, air pollution, malfunctioning equipment, inadequate lighting, and poor equipment design. Inefficient work methods due to poor equipment design subject dental professionals to considerable fatigue, which in turn makes them more susceptible to psychological stress.

Many dentists and dental auxiliaries feel that everyone must like them and that they must like everyone. In the event that dental professionals experience lack of acceptance or even outright rejection by some people, they might (and often do) feel that they are not okay. This rejection indicates (to them) imperfection. Being confronted with the reality of interpersonal relationships—and in fact, their own limitations—becomes stressful.

In their study of "Dental Family Stress and Coping Patterns," Nevin and Sampson state, "Inability to accept limitations is a problem common to many people in dentistry and a great source of stress for many dentists in particular." In fact, Nevin and Sampson found that 64% of the dentists they surveyed said that perfectionism was their greatest source of stress.

The entire team is often negatively affected when a patient experiences discomfort. The team's entire purpose is to eliminate discomfort and to gain and insure health. For patients to fear them as pain inflictors—however misperceived—is often debilitating.

EXTERNAL OR NON-JOB-RELATED FACTORS

1. Family relationships
2. Commuting time and distance
3. Economic factors
4. Demands placed on an individual by the community
5. Social activities

Family relationships are often the most significant external factors leading to excessive stress. Sometimes families receive the stress inflicted by the dental professional, or sometimes it is the family itself that is the stress producer for the dentist or for the dental auxiliary. Stress in dentistry can have a strong and direct effect on the physical well-being of both the dentist and the spouse. When the dentist is feeling upset or stressed out, the spouse is likely to be the same.

The stressors that seem to emerge as the most intense for dental families are finance and business strains (changes in the financial conditions of the practice, thus affecting available funds for family use).

Dental families experience stress that arises from both the dental practice and from family-related issues. Stable dental families exercise effort to maintain a sense of balance through effective coping skills and through family resources. The results of Nevin and Sampson's study of dental-family stress and coping patterns indicated that

> *"strong coping patterns resulted when dental professionals and spouses maintained a balance of time and responsibility, satisfaction in work and family activity,* regular communication, *sharing of decision making, good physical health, and the inclusion of an active exercise program within multiple demands of their time."*

PERSONALITY CHARACTERISTICS

Personality characteristics play a significant part in how an individual deals with stressors. As was previously indicated, certain stressors generate different actions and reactions from different people. One of the main reasons behind this is that different people wear different character makeups.

Most generally, behavior types fall under two distinct categories. Type A behavior is characterized by excessive aggressiveness, time urgency, restlessness, hostility, and tenseness. Type A dental professionals have been found to be behind schedule more frequently then Type B dental professionals.

Type B characteristics are flexibility, ability to relax, control, and acceptance. Type B individuals feel in control of their lives in comparison to Type A personalities, who feel that others control their lives, that external forces have a stronger effect on them than internal forces.

EFFECTS OF STRESS

Stress affects people both physiologically and psychologically. Psychologically, stress "takes the form of a subjective sense of discomfort and distress, distortions or thinking, decreased performance, and indecisiveness," according to Hans Selye. These psychological effects may lead to:

1. Lack of desire to go to work
2. Absenteeism
3. Poor or altered performance
4. Difficulty in getting along with peers/co-workers

Physiologically, the stress response consists of a coordinated set of bodily responses signaled by a release of hormones and sympathetic nervous system stimulation.

Physiological effects of stress may lead to:

1. High blood pressure and other coronary problems
2. Ulcers and other digestive problems
3. Illness such as colds and flu
4. Headaches and migraines
5. Sleeping disorders

Physiological and psychological consequences of stress, although separate, are nevertheless interrelated. Illnesses resulting from stress will definitely have an effect on job performance, presence on the job, attitude on the job. And conversely, poor attitude about work, conflicting interpersonal relationships among co-workers, outside negative influences, such as financial difficulties and marital problems, will lead to high blood pressure and other above mentioned physiological diseases.

Stress left unbridled can become detrimental to one's mental and physical health. A common thread runs through the lives of most people who fall prey to the potentially devastating effects of uncontrolled stress. This thread is that most of the stress could have been tamed—could have been controlled. Total mismanagement of the stress has led to illnesses, misery, and even death.

The correctable nature of much of this mismanaged illness is not to say that the results are not debilitating. Not in the least. The devastating effects are concrete—no doubt about it.

1. Physically: people get sick. They often die earlier.
2. Financially: people do not perform at their best.
3. Emotionally: people are not at their peak and do not handle their relationships (or themselves) well.
4. Spiritually: people are often depressed and lay blame. They choose *not* to develop a close spiritual relationship.

In spite of books, tapes, and courses on the subject of stress management and personal development, many people do not choose to manage themselves correctly. This is reflected in illness and death statistics, in the quality of people's daily lives, and ultimately in the enormous loss of profits in business—tens of billions of dollars per year.

Signs of poor self-management include:

1. Low energy
2. Decreased productivity
3. Lowered self-image/self-esteem
4. Poor decision making
5. More mistakes
6. Accidents
7. Illness
8. Tardiness and/or absenteeism
9. Burnout
10. Problems with personal relationships at home and at work
11. Poor financial management
12. Shortened longevity

Those people who choose not to take personal responsibility for the control of stress in their lives, but who blame their job, other people, or circumstances for their problems are selecting a self-destructive route. When days get full and hectic, the thing that usually falls by the wayside is time spent for personal management: exercise, eating properly, relaxation, time with kids, spouse, or self.

You are responsible to yourself. You are the manager of yourself. If you think you are special, you are! But not so special that you are immune to the risks of poor personal management. Workaholics, alcoholics, drug abusers, cigarette smokers, obese people think that problems are going to happen to someone else. Wrong! You must run your body/mind with the same expertise that you run your practice.

Again, stress cannot be eliminated—nor would you want to eliminate it. The challenge is to learn how to control or manage your personal response to stress. Those who choose to control their stress by actively and constructively managing it can learn to harness stress and can turn this potentially harmful force into high-powered energy.

Dr. Hans Selye claims,

> *"The same stress which makes one person sick can be an invigorating experience for another. It is through the General Adaptation Syndrome that our various internal organs help us both to adjust to the constant changes which occur in and around us and to navigate a steady course toward whatever we consider a worthwhile goal. Through the constant interplay between mind/body, man has the power to influence his adaptation. We cannot avoid stress but we can control 'distress' and keep its damaging effects to a minimum."*

The secret to happiness lies in the successful adaptation to ever-changing conditions. Inability or unwillingness to adapt leads to disease and unhappiness. Effective adaptation can only occur if the pleasure of achievements far outweighs the pain of change. Many people refuse to adapt or to change or to relearn. Thus any adaptation becomes more complicated, more difficult.

THE MANAGEMENT OF STRESS

We have defined stress. We have identified sources of stress inside and outside of the dental environment. We have briefly described personality characteristics that may affect a person's ability to cope with stressors. And we have detailed some of the effects of stress on the physical and mental well-being of individuals. How does a person control this inevitable force in the daily dental workplace? How does one turn a potentially negative influence into a productive, positive motivator?

The skill of stress management *can* be learned! Identifying a stressor and your own ability or inability to handle a stressor is the first step in stress control. Awareness is the first step toward growth or change.

Let's identify some of the specific stressors that may be leading to your stress. Here are some of the things that cause the most stress in the work environment:

1. Too much work
2. Too little work
3. Poor direction at work
4. Workaholism
5. Not being able to stay on top of technical advancements
6. Poor communication
7. Special stresses
    - Challenges for women in the workplace
    - Single parents—dad or mom
8. Travel and catching up when one gets back home
9. Inability to relax
10. Substance abuse

PROVEN WAYS TO REDUCE AND CONTROL STRESS

Without question, one of the most effective ways to control stress is through regular exercise. Studies have proven that during exercise, tranquilizing chemicals called endorphins are released into the brain. This natural chemical brings about a pleasurable sensation. Physicians who study the positive effects of exercise on the mind/body encourage a minimum of 30 minutes of vigorous exercise three to four times per week, with five times per week being even more effective. The key to the success of an exercise program is not to see how exhausted you can get, but to exercise your heart on a regular, cyclical basis.

Consider these exercise programs:

1. Walking

2. Swimming

3. Jogging

4. Bicycling

5. Cross-country skiing

Always consult your physician before embarking on an exercise program. Warm up and stretch before beginning any exercise session. Don't overdo. Increase the amount of exercise a little bit at a time. Cool down at the completion of each exercise session.

In addition to identifying and confronting stressful situations and including a regular exercise program in your life, research has proven that simple relaxation techniques when practiced faithfully can and will reduce the negative effects of stress.

Commonsense tips on managing stress are:

1. Get enough sleep: seven hours per night are recommended for an adult.

2. Work out your anger in some useful activity, not on a person!

3. Talk out your worries with a concerned confidante.

4. Manage your time wisely. Make a list of the things you need to do each day.

5. Keep your routine orderly and efficient. Don't try to do everything at once.

6. List your necessary and desired goals and then prioritize them! Stick with number one until you get it done—then go to number two.

7. Eat properly. "You are what you eat!"

8. Feed your mind positive "food." As Earl Nightengale says, "You become what you think about!"

9. Take breaks during your day. If nothing else, take a stand-up-and-stretch break. Take a break at lunch. Have a light snack, get a breath of fresh air, and stretch. You'll be much better for that 1:00 patient, for that 4:00 patient, and for the family that awaits your arrival at home.

10. Don't kid yourself into thinking that you can eliminate or control stress with substance abuses. Make sure that you moderate or eliminate your intake of alcohol, caffeine, nicotine, barbiturates/tranquilizers. Control your stress naturally with a commitment to health of both the mind and the body.

Mental Fitness

Productivity and profitability depend on the mental and physical fitness of the individual members of the team. Here are some way to ensure both:

1. Set realistic goals. Positive thinking that is not supported by planned action is simply wishful thinking. Learn the art and science of goal setting (Chapter 2). Study and become motivated by the actual results that people are seeing in their lives and in their practices as a result of setting goals and disciplining themselves to bring those goals into realities. Realistic goals and appropriately related behavior leads to achievement.

2. A well-prepared person with a vision and specifically written goals has a focused direction. This will affect confidence and will make way for great accomplishment. This relieves frustration, stress, and burnout.

3. Create mental fitness in the workplace. Do this by continuing to learn the newest developments in your field, by adopting a positive approach, and by keeping the right side of the brain (the creative side) actively seeking and receiving creative solutions. Then you can actually reduce the number of bad days.

4. Commit to getting and maintaining organization in all parts of

your practice. Have excellent management systems in place and have talented, well-trained people engineering those systems.

5. Study and learn excellent communication skills that can be used to more effectively listen and to speak. Access the skill and knowledge of confrontation so that conflicts can be resolved constructively. Communicate effectively with other team members and with patients.

6. The person who works at having a positive attitude and surrounds him/herself with positive people will always outshine the person who wallows in negativity, despair, and gloom.

7. Invest in courses for employees on stress management. This will come back to the practice as a fabulous investment. There will be less sick leave and much greater productivity.

8. Pay attention to prevention of stress-related problems. Don't fail to realize that the results of uncontrolled stress can be harmful. Incompetent management of human resources—team members—is one of the greatest single wastes in the cost of doing business today.

Relax

People tend to blame stress for driving them to the brink of loss of control. They blame their jobs for this stress. This does not need to happen at all. Each person chooses how they will respond or adapt to the stresses that are a part of modern-day dentistry. Take care of yourself as a good manager of your greatest gift and asset: *You!* Learn to relax. Learn to control your bodily responses.

Dr. Herbert Benson in his well-known best seller, *The Relaxation Response,* states:

> *"The relaxation response is the inborn capacity of the body to enter a special state characterized by lowered heart rate, decreased rate of breathing, lowered blood pressure, slower brain waves, and an overall reduction of the speed of metabolism. These changes produced by the relaxation response counteract the harmful effects and uncomfortable feelings of stress."*

The sidebar is a synopsis of Dr. Benson's *Relaxation Response*. The results are phenomenal and have proven beneficial. The few moments spent in concentrated relaxation and control are worth the results and far outweigh the opposite: lost control!

Dr. Benson recommends that the relaxation response be practiced for 15–20 minutes two times per day, preferably morning and evening. But once mastered, the technique can be performed at any time for even a few minutes with positive effects. Through the relaxation response, such physical changes as the following can be induced:

1. Lowering of heart and breathing rates
2. Lowering of blood pressure
3. Improvement of the body's immune mechanisms
4. Increased levels of killer T white blood cells in the bloodstream
5. Elevation of circulating levels of endorphins (making pain tolerance greater)
6. Control of headaches, even migraines

THE RELAXATION RESPONSE

1. Sit quietly in a comfortable position, and loosen any tight clothing around your waist (if possible).
2. Close your eyes.
3. Deeply relax all your muscles, starting with your feet and moving up to your head. Keep your muscles relaxed.
4. Breathe through your nose. Become aware of your breathing. As you breathe out, repeat a word or brief phrase over and over, such as "one."

   Breathe in, then out, and say "one." Breathe in, then out, and say, "one." Breathe easily and naturally—through your nose. Breathe into your stomach. Breathe out through your mouth. Sometimes your breaths will be deeper than others: that's okay. Continue.
5. Continue for 15–20 minutes. You may open your eyes to check

the time, but do not use an alarm or you will subconsciously prepare for the ringing of the alarm.

6. When you finish, sit up quietly for several minutes. Keep your eyes closed, then after a minute or two, open them. Do not stand up immediately.

Do not become distressed if your mind wanders. It will do this. Just recognize the thought and quietly bring your mind back to concentrating on the breathing and to the repeating of your word or phrase. Once you have mastered the skill, you can gain control of yourself and of your stress by practicing the skill for even a few minutes when a brief break is available.

Other Effective Relaxation Techniques

Hobbies: Do something you really enjoy, and do it on a regular basis for at least one-half hour per day. These activities can provide a creative outlet totally different from the activities of your working day. In addition to hobbies such as sewing, reading, music, carpentry, gardening, cooking, and sports, look into adult education programs, civic and church activities. The change will reduce stress and fatigue while refreshing the creative part of your mind.

Meditation: Similar to the Relaxation Response, the various forms of meditation have proven effective in reducing stress through the power of the mind. These techniques are not necessarily tied to any specific religious belief.

Biofeedback: Special medical instruments handled by a skilled instructor can help a person control their reaction to specific situations. Again, the mind is trained to control the bodily responses.

Hypnosis: This technique can be used to bring about a relaxed, stress-free state in a person. It can also be used to break stress-related habits such as overeating, smoking, alcohol, and drug abuse.

Visualization: Taking an imaginary trip to a pleasant trip from your past—or a hoped-for trip in the future can relax the mind and the body. Close your eyes, take a deep breath, and for a brief 5–10 minutes imagine the details of your dream location. Visualize the scene in detail. Feel the warm sun on your skin. Hear the waves crashing on the shore. Taste the water. In this brief "trip" you can relax into a vacation-like state.

The key to surviving and thriving on stress is control. Learn to ignore what you can't control and learn to control what you can.

1. Give those with whom you work your trust and confidence. Be patient and understanding. Be honest. Be committed.
2. Give those you love more of your time and focused attention.
3. Give yourself and the people with whom you interact the benefits of effective communication skills. Care enough to communicate.
4. Give yourself rest, proper nutrition, mental and physical exercise, relaxation, play, dreams, and goals.
5. Give yourself a lifetime of purpose by pursuing work that you love. If you retire, turn your energy to another type of work that fulfills a purpose.
6. Give yourself a sense of challenge each and every day.

TWELVE STEPS TO SUCCESSFUL STRESS CONTROL

1. Write your individual and team goals: together, focus on where you are, where you are going, and how you intend to get there.
2. Prioritize your goals.
3. Practice good time management.
4. Exercise on a regular basis for physical and emotional well-being.
5. Practice proven relaxation techniques.
6. Feed your body properly.

7. Feed your mind positively.
8. Stimulate your creative centers with hobbies and outside activities.
9. Communicate effectively. Face stressful situations head-on with constructive confrontation skills.
10. Organize every aspect, every system within your practice. Pay attention to detail. Do things right the first time. Eliminate chaos. Chaos breeds stress.
11. Eliminate the things in your life that are not working and focus on and give attention to the things in your life that are working. Get control of your life instead of letting your life control you. All of life is a matter of choice.
12. Strive for achieving a balance in your life: a balance between love, work, worship, and play.

(Aristotle/L. D. Pankey)

IN SUMMARY

When dental professionals have been exposed to one or more stressors, there are two reactions that might take place: (1) psychological reaction and (2) physiological reaction. Psychological reaction to stress has been found to increase a person's anxiety, has led to depression, and may cause decreasing job satisfaction. The behavioral consequences that occur as a result of this psychological reaction to stress may be:

1. Reduced job performance
2. Absenteeism
3. Tardiness

4. Job turnover

5. Interpersonal conflict

Physiological reaction to stress leads to such physical problems as:

1. High blood pressure

2. Gastric problems

3. Headaches and migraines

4. Colds and flu

5. Difficulty sleeping

These two reactions to stressors, psychological and physiological, are interrelated in that psychological disorders or a person's inadequate ability to cope with daily dental stresses can cause or lead to physiological disease. On the other hand, physiologic reaction and disease will affect a person's productivity on the job, adding to absenteeism, tardiness, and job turnover.

Stressful situations in the dental environment will never be eliminated, but they can be controlled. Developing coping strategies and learning how to deal constructively with the daily activities of the dental profession and in daily living are a *must* if stressors are not to take a negative toll on dental professionals.

CHAPTER 14

AND IN CONCLUSION

"If you have definitely determined what you want and have fixed a goal for yourself, then consider yourself extremely fortunate for you have taken the first step that will lead to your success. As long as you hold on to that mental picture of your idea and begin to develop it with action, nothing can stop you from succeeding. The subconscious mind never fails to obey any order given to it clearly and emphatically."

CLAUDE BRISTOL

I made the statement at the beginning of the book that I could not start a lecture, a consultation, or this book without discussing the benefits of and the positive results of goal setting. I meant that with all of my heart because I have been such a committed goal setter for the majority of my adult life. One of the goals that I set many years ago was to write this book and to become involved with a publishing company that believed in me and in my work. This book is not only a dream come true, but a goal accomplished.

What a joy it has been to put together a summation of my thoughts, my learning, my experience in the dental industry. Throughout the years of my work and throughout the months of the writing of this book, I have become even more convinced that communication *is* the bottom line to success in the dental practice. Whether communicating with members of the team or with patients, personal, professional, and financial success is made or broken by the manner in which communication takes place.

Becoming more attentive and accurate in your listening skills alerts you to the perceived needs of others. With this powerful information, you can be about the business of answering those needs. You expand your people skills and bond with your patients by developing a common thread of respect and mutuality.

By speaking and expressing yourself clearly and effectively you can access results that you could not otherwise achieve. You will keep the lines of communication open. Closing the doors to communication will not get you where you want to go. Be ever aware of Paul Harvey's classic statement: "It's not what you say, it's how you say it!"

Conflicts *will* arise. Understanding the differences between people and knowing how to function with—rather than against—other people moves you forward. Otherwise, relationships go nowhere or backwards. When conflicts do arise, you can now face them positively, because you have the armamentarium to turn conflicting situations into caring relationships.

The American Dental Association tells us that for every 10 years that a doctor practices dentistry that he/she has approximately $1,000,000 worth of dentistry diagnosed but left untreated. In other words, for every 10 years that a doctor practices dentistry, he/she has about $1,000,000 worth of dentistry still sitting in the charts waiting to be done.

With excellent communication skills that become the underlying foundation of successful case presentations, more people will say yes to your recommendations. Personal and professional fulfillment will result. You will be providing the type of dental care in which you believe and the type of care you have spent a lifetime studying and perfecting. Rather than doing patchwork dentistry, you will be providing more comprehensive dental care.

There *will be* objections to face, but you will be able to do so with a new attitude, an attitude that says, "Thanks for the objection. Now I know that you, Mrs. Patient, are interested in the proposal and are seeking answers to questions. If you didn't have those questions or those objections you would be telling me that you are not interested. Not only will I look forward to your objections, but I will be able to handle them effectively."

Patients who become a part of your practice will not be walking out the door not receiving care. They will be walking out the door healthier, more beautiful, and happier than they were before they walked in. They will walk out happier, but they will come back, because as people professionals, you will have nurtured the relationship so that they will not only come to you for treatment, but they will stay with you for the duration. And as happy, satisfied clients, they will refer other people to you that will receive the same, consistent care each and every time they enter your practice.

When Dr. Saddoris first "discovered" me, he was the reigning president of the American Dental Association. Over lunch one beautiful afternoon in his lovely city of Tulsa, Oklahoma, he asked me to explain the process of my consultations with doctors and their teams. I showed him a graph outlining the 20 areas of the dental practice that we streamline when we go through a comprehensive consultation experience.

He asked, "Which of these 20 areas of concentration do doctors want and need the most, Cathy?"

Without hesitation, I answered, "Communication skills. This is usually the number one area of need and the number one area of requested help."

Therefore, it is tremendously satisfying to be putting into written form the ideas, concepts, strategies, and skills that I teach every day to dental professionals through lecture and through consultation. As a "forever" student myself, I will constantly be learning new and better ways to communicate. The day I stop learning is the day I need to stop teaching. This will never happen, because I find that every opportunity I have to enhance my own communication skills makes my personal and professional life richer, smoother, and more successful.

This is my hope for you: that by studying this book on communication skills, that you will be richer—personally, professionally, and financially. I truly believe—I know—that if you will apply the skills held within this manuscript that you will grow and develop to a new level of confidence, and as a result, you will accelerate in all areas of your life.

Call me with your successes. Call me when you realize that: Great communication *does* equal great production.

Appendix A

Controlling the Financial Arrangements in *Your* Practice

By Cathy Jameson, M.A.

As I consult across the country, I rarely go into a practice where a written financial policy is in place. That is one of the first projects we commit to: the designing, the writing, and the implementation of a financial policy.

In this article, five steps that need to be taken to implement a financial policy will be outlined. Then a financial policy that will work for most any practice will be described.

Step 1. Set and write the policy. Making arbitrary, random decisions about *who's* going to pay and *how* they will pay will lead to financial chaos for your practice. Loss of control of accounts receivable can result. Problems and misunderstandings from patients in regard to their financial responsibility can arise.

Historically, many doctors have feared setting a firm financial policy because they thought that patients would leave the practice if they weren't lenient with the financial aspect of treatment. The financial policy I am going to recommend will allow for necessary flexibility but will also provide necessary firmness for practice solidity.

Step 2. Decide, as a team, what financial options you are going to make available. "If a person is allowed to be a part of a decision-making process, that person will be more likely to buy into the decision." Abiding by this tried and true management principle, you will see that if the entire team understands the policy options and decides—together—which options are applicable and appropriate to your practice, they will support the policy.

If team members question the financial policy, that lack of confidence will come through loud and clear to patients. *Their* insecurity will lead to *patient* insecurity.

Step 3. Financial arrangements *must* be made before any treatment is provided. Patients have made it clear that they don't like not knowing how much the treatment will cost before it's done. We must listen to our consumers/patients and respond positively to what they are requesting. They are saying, "Let me know in advance what my financial responsibility will be."

Step 4. Designate a specific person on the team to administer financial arrangements. This person (probably the business manager) would review the treatment plan that the doctor has discussed, define the fee, analyze expected insurance (if applicable), and determine the financial option that is acceptable to the patient and to the practice. This financial arrangement would be written so that neither party would forget or become confused about the agreement.

In our own practice, the business manager joins the doctor and the patient for the consultation appointment. The doctor asks the patient for permission. He says, "Mrs. Jones, I've asked Pam, my financial coordinator, to join us today for our consultation. Pam handles the financial arrangements for our patients, and she schedules the appointments. Therefore, I felt it was important for her to hear the recommendations that I'm making for you. Are you comfortable with that?" No one has ever said no.

The doctor presents his recommendations, answers treatment questions, asks for the commitment to go ahead with treatment, and then, when there are no other questions except financing, he excuses himself. Pam takes over. She reconfirms the treatment, defines the total investment, the investment per appointment, and the financial options we have available. Together, she and the patient clarify all financial responsibilities. She writes this down. Then she schedules the first appointment.

Step 5. Do your best to create a time and a place for private consultation, both clinical and financial. The oral cavity is an intimate zone of the body. So is the pocketbook! To discuss treatment recommendations or financing in public is uncomfortable and awkward.

Find the place! Create the appropriate environment! It doesn't have to be fancy. It just has to *work!* Schedule the time for this consultation. This consultation is as critical as any phase of your treatment!

1. Preplan what you're going to do and who's going to do what.

2. Practice the skills.

3. Perfect the skills.

4. Perform the arrangements.

Those are Tom Hopkins' four Ps of professionalism: preplan, practice, perfect, perform! Good advice!

A WORKABLE FINANCIAL POLICY

The financial policy that I recommend is as follows:

Option 1. Five percent accounting reduction for advance payment of treatment in full.

My friend, Dr. Mitch Cantor, a periodontist in Southampton, NY uses the following verbal skills to present this option:

DOCTOR: Mrs. Smith, many of our patients have chosen to take advantage of the opportunity to have their fee reduced.

MRS. SMITH: Oh, really! How do they do that?

DOCTOR: When you take care of your financial responsibility in advance of the treatment, we will reduce your fee by 5%. Would this be of interest to you?

Great verbal skills! He finds that *many* patients take advantage of this option. You are dollars ahead to offer this reduction vs. carrying any accounts on your own books.

Note: If this option is used with an insurance patient on their private-pay portion of the treatment, you must stamp on the insurance claim, "This person has received an accounting reduction of 5%."

Option 2. Payment by appointment. Great option! You *must* have excellent, written treatment plans that give the business manager the necessary information to determine what will be done at each appointment. Then, and only then, can she inform a patient of their financial responsibility per appointment. (Remember, she needs to do this in advance of each appointment!)

Option 3. Insurance on assignment. A recent ADA survey asked people across the entire country the following question, "If you needed to make a one-time dental purchase of $500, could you?" Seventy-seven percent of the people who responded said no.

Understanding this critical fact, I usually recommend taking insurance on assignment. If you are in a high socioeconomic area, if your patients can afford to pay for their care and wait for the insurance company to reimburse them, directly, then, by all means, do just that!

However, for the vast majority of practices, not accepting assignment would eliminate a great many patients and a great deal of dentistry.

Insurance management is not easy. You must have an effective, efficient system in place. But the financial reward and the service to your patients is worth the efforts.

Option 4. MC/VISA/Discover Bank Cards. These wonderful financing vehicles have not historically been used much to finance dentistry. In fact, less than 6% of the dentistry in the United States is done on MC/VISA.

However, this trend is on an upward swing.

These are *very* desirable options. Don't worry about the service charge. Running your own credit business is financially devastating and inadequate at best.

You must *ask* for these cards. Don't just hang up a sign and think people will respond. *You* have to mention and *encourage* the availability of the service.

Option 5. For any long-term or extended payments, a health-care financing program (or health-care credit card).

These programs have become an asset to the dental industry by providing convenient financing for comprehensive or immediate care.

Using these programs gets you, the doctor, out of the banking business while still allowing patients the opportunity to spread the payment of treatment out over several months. Monthly payments would be small and comfortable to the family budget. Call these companies. Write to them. Gather the information that details the individual programs. Become involved, *then learn how to use the program as a service to your patients and to you!*

You will find more people saying yes to your treatment recommendations when you defuse that awesome and foreboding fear of cost.

This financial policy takes care of most people:

1. Those who can afford to pay up-front.
2. Those who can pay as treatment proceeds.
3. Those who need a long time to pay with minimum monthly investments.

We follow this *exact* policy and have for about 10 years. We have *no* accounts receivable except a two to four week turnaround on insurance. Our productivity is higher than ever and grows every year. Along with that production comes increased profits because we've eliminated the cost of running a credit business within our practice.

Options are available today that will allow you to get out of the banking business and to spend your valuable time and hard-earned money on providing dental care and on building your practice.

Steps have been outlined for organizing a financial program for your practice. A financial policy has been suggested. You need and want to collect that which you produce and you want to meet the financial needs of your patients—a win/win situation.

You have the guidelines. Now, just *do* it!

Bibliography

Adams, Linda. *Be Your Best,* New York: The Putnam Publishing Group, 1989.

Adler, Mortimer. *How to Speak, How to Listen,* New York: Macmillan Publishing Co., 1983.

Ash, Mary Kay. *On People Management,* New York: Harper and Row, 1984.

Benson, M.D., Herbert. *The Relaxation Response,* New York: Morrow and Co., 1975.

Blanchard, Ph.D., Kenneth, and Johnson, M.D., Spencer. *The One-Minute Manager,* New York: Berkeley Books, 1982.

Blanchard, Kenneth. *Raving Fans,* New York: Morrow and Co., 1993.

Caplan, D.D.S., Carl. *Dental Practice Management Encyclopedia,* Tulsa: PennWell Books, 1985.

Christen, A.G. "Stress and Distress in Dental Practice," *Occupational Hazards in Dentistry,* Chicago: Yearbook Medical Publishers, Inc., 1984.

Cinotti, D.D.S., William; Grieder, D.D.S., Arthur; and Heckel, Ph.D., Robert. *Applied Psychology in Dentistry,* St. Louis: C.V. Mosby, 1964.

Couture, Ph.D., Gary. *The Quantum Breakthrough in Communication,* Newport Beach, CA: Institute for Advanced Educational Research. Audiotape.

Covey, Stephen R. *Daily Reflections for Highly Effective People,* New York: Simon & Schuster, Inc., 1994.

_______. *Principle Centered Leadership,* New York: Summit Books, 1991.

Dahl, Dan, and Sykes, Randolph. *Charting your Goals,* New York: Harper and Row, 1983.

Dworkin, D.D.S., Samual; Ferrence, Ph.D., Thomas; and Gidden, Ph.D., Donald. *Behavioral Science and Dental Practice,* St. Louis: C.V. Mosby, 1978.

Eliot, M.D., Robert. *From Stress To Strength,* New York: Bantam Books, 1994.

Folger, Joseph, and Poole, Marshall. *Working Through Conflict,* Glenview, IL: Scott, Foresman, 1982.

Foreman, Ed. *Happy, Healthy, and Terrific,* Dallas: Executive Development Systems, 1988.

_______. *Laughing, Loving, and Living,* Dallas: Executive Development Systems., 1982.

Gordon, Ph.D., Thomas. *Leader Effectiveness Training,* New York: Peter Wyden Publishers, 1977.

_______. *Parent Effectiveness Training,* New York: Peter Wyden Publishers, 1974.

Hanson, M.D., Peter G. *The Joy of Stress,* Canada: Hanson Stress Management Organization, Inc., 1985.

_______. *Stress for Success,* Canada: Hanson Stress Management Organization, 1989.

Hendrix, Ph.D., William. "Dental Stress and Assessment Questionnaire," *Dental Clinics of North America* (Philadelphia: W.B. Saunders) Supplement 30:4 (October 1986):S1–S10.

Hopkins, Tom. *How to Master the Art of Selling,* Scottsdale, AZ: Warner Books, 1982.

_______. *The Official Guide to Success,* Scottsdale, Arizona: Tom Hopkins International, Inc., 1982.

Jackson, Eric. "Stress Management and Personal Satisfaction in Dental Practice," *Dental Clinics of North America* (Philadelphia: W.B. Saunders) Volume 21:3 (July 1977):559–576.

Karrass, Gary. *Negotiate to Close,* New York: Simon & Schuster, 1985.

Katz, D.D.S., Clifford. "Stress Factors Operating in the Dental Office Work Environment," *Dental Clinics of North America* (Philadelphia: W.B. Saunders) Supplement 30:4 (October 1986):S29–S36.

LaHaye, Tim. *Spirit-Controlled Temperament,* San Francisco: Post, Inc., 1966.

LeBoeuf, Michael. *The Greatest Management Principle,* Chicago: Nightingale-Conant, 1986.

Littauer, Florence. *Personality Plus,* Old Tappan, NJ: Power Books, 1983.

Miller, Ph.D., Lyle, and Smith, Ph.D., Alma Dell. *The Stress Solution,* New York: Pocket Books, 1993.

Murphy, Kevin J. *Effective Listening: Your Key to Career Success,* New York: Bantam Books, 1989.

Nevin, Ph.D., Robert, and Sampson, M.S., Vicki. "Dental Family Stress and Coping Patterns," *Dental Clinics of North America* (Philadelphia: W.B. Saunders) Supplement 30:4 (October 1986):117–131.

Pillow, William. *Communication with Patients*, Indianapolis: Eli Lilly and Co., 1985.

Selye, M.D., Hans. *The Stress of Life,* New York: McGraw-Hill, 1978.

_______. *Stress Without Distress,* New York: Signet, 1974.

Solomon, Muriel. *Working With Difficult People,* Englewood Cliffs, NJ: Prentice Hall, 1990.

Swogger, M.D., Glenn. "The Type A Personality, Overwork, and Career Burnout," *Dental Clinics of North America* (Philadelphia: W.B. Saunders) Supplement 30:4 (October 1986):37–44.

Verderber, Rudolph. *Communicate*, Belmont, CA: Wadsworth Publishing Company, 1987.

Willingham, Ron. *The Best Seller—The New Psychology of Selling and Persuading People,* Englewood Cliffs, NJ: Prentice-Hall, 1984.

Ziglar, Zig. *See You At The Top,* Gretna: Pelican Publishers, 1988.

_______. *Ziglar On Selling,* Nashville: Oliver-Nelson Books, 1991.

INDEX

A

B

E

F

G

H

I

J

K

L

M

N

O

P

Q

R

S

T

V

W

Jameson Management Group

THE COMPANY'S MISSION AND VISION

The mission of Jameson Management Group is to serve the health-care industry as a consulting and lecturing service that provides both business and personnel management. The following is a statement of mission and purpose for Cathy Jameson and Associates and Jameson Management Group Consulting Services:

1. To make a positive difference in the lives of the professionals whom we have the privilege to serve.
2. To teach practice management with a heart.
3. To teach practical skills of business and personnel management.
4. To facilitate excellent relationships between and among team members—via the teaching of effective communication skills.
5. To help team members understand these communication skills and be able to serve patients better because of productive communication.
6. To integrate business systems that will lead the way to productivity and profitability for the practitioners and the entire team.
7. To teach systems that are time efficient, cost efficient, and will control stress.

Cathy Jameson provides in-office consulting services and seminars through her company, Jameson Management Group, Inc., located in Oklahoma. For more information, please call (405) 369-2501 or (405) 369-5555. Correspondence and inquiries can be sent to:

Jameson Management Group, Inc.
P.O. Box 488
Davis, OK 73030-0488